信息科学技术学术著作丛书

开放式遥感数据处理软件平台 OpenRS 的设计与实现

江万寿 吕维 张靖 著

科学出版社
北京

内容简介

本书简要地介绍开放式遥感数据处理软件平台 OpenRS 的架构设计和实现方法。全书共 10 章，包括应用无关的插件平台、面向遥感影像处理的对象体系和基础模块、面向插件开发的可扩展界面、属性化的参数界面、基于可执行对象的处理流程、分布式平行处理框架和一键式服务包装等核心内容。OpenRS 是一个开放的平台，也是一个开源的平台，本书的内容是与其源代码紧密相关的，目的是对开放式遥感数据处理软件平台 OpenRS 做一个比较系统的介绍，为促进国产遥感开源软件的发展贡献一份力量。

本书可供摄影测量、遥感数据处理相关软件的系统设计人员、开发人员参考，也可供对开源软件感兴趣的爱好者研究和学习。

图书在版编目(CIP)数据

开放式遥感数据处理软件平台 OpenRS 的设计与实现/江万寿，眭维，张靖著. —北京：科学出版社，2016

（信息科学技术学术著作丛书）

ISBN 978-7-03-048997-5

Ⅰ. ①开… Ⅱ. ①江… ②眭… ③张… Ⅲ. ①遥感数据-数据处理软件 Ⅳ. ①TP751.1

中国版本图书馆 CIP 数据核字(2016)第 141185 号

责任编辑：张艳芬 罗 娟 / 责任校对：郭瑞芝
责任印制：徐晓晨 / 封面设计：左 讯

科学出版社 出版
北京东黄城根北街 16 号
邮政编码：100717
http://www.sciencep.com

北京京华虎彩印刷有限公司 印刷
科学出版社发行 各地新华书店经销

*

2016 年 8 月第 一 版 开本：720×1000 1/16
2017 年 1 月第二次印刷 印张：21
字数：407 000
定价：128.00 元
(如有印装质量问题，我社负责调换)

《信息科学技术学术著作丛书》序

21世纪是信息科学技术发生深刻变革的时代,一场以网络科学、高性能计算和仿真、智能科学、计算思维为特征的信息科学革命正在兴起。信息科学技术正在逐步融入各个应用领域并与生物、纳米、认知等交织在一起,悄然改变着我们的生活方式。信息科学技术已经成为人类社会进步过程中发展最快、交叉渗透性最强、应用面最广的关键技术。

如何进一步推动我国信息科学技术的研究与发展;如何将信息技术发展的新理论、新方法与研究成果转化为社会发展的新动力;如何抓住信息技术深刻发展变革的机遇,提升我国自主创新和可持续发展的能力?这些问题的解答都离不开我国科技工作者和工程技术人员的求索和艰辛付出。为这些科技工作者和工程技术人员提供一个良好的出版环境和平台,将这些科技成就迅速转化为智力成果,将对我国信息科学技术的发展起到重要的推动作用。

《信息科学技术学术著作丛书》是科学出版社在广泛征求专家意见的基础上,经过长期考察、反复论证之后组织出版的。这套丛书旨在传播网络科学和未来网络技术,微电子、光电子和量子信息技术、超级计算机、软件和信息存储技术,数据知识化和基于知识处理的未来信息服务业,低成本信息化和用信息技术提升传统产业,智能与认知科学、生物信息学、社会信息学等前沿交叉科学,信息科学基础理论,信息安全等几个未来信息科学技术重点发展领域的优秀科研成果。丛书力争起点高、内容新、导向性强,具有一定的原创性;体现出科学出版社"高层次、高质量、高水平"的特色和"严肃、严密、严格"的优良作风。

希望这套丛书的出版,能为我国信息科学技术的发展、创新和突破带来一些启迪和帮助。同时,欢迎广大读者提出好的建议,以促进和完善丛书的出版工作。

中国工程院院士
原中国科学院计算技术研究所所长

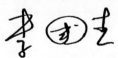

前　　言

"十一五"期间,国家"863"计划对地观测领域重点项目设立了"遥感软件体系架构及标准规范研究"课题(2007AA120203)。在该课题的支持下,武汉大学课题组设计开发了开放式遥感数据处理软件平台 OpenRS,旨在实现可扩展、可伸缩、可定制的遥感数据处理目标。该平台是一个半开源的项目。为了更好地推动该平台的应用,使更多的遥感数据处理算法开发人员能够共同开发一个实用的遥感数据处理软件系统,作者撰写了本书,以对开放式遥感软件平台 OpenRS 的设计与实现思路进行较全面的总结,为平台的推广应用提供一个比较系统的参考。

本书共 10 章。第 1 章主要介绍设计背景和设计目标,第 2 章介绍总体架构设计和实现思路,第 3 章介绍插件系统设计与实现,第 4 章介绍遥感数据处理所需的基本模块,第 5 章在第 3 章的基础上介绍基于 MFC 和 BCG 界面库的界面扩展设计,第 6 章介绍桌面集成环境设计与实现,第 7 章介绍处理流程设计与实现,第 8 章介绍分布式并行处理设计与实现,第 9 章介绍网络服务包装与嵌入应用,第 10 章介绍典型的插件开发实践。其中第 1 章和第 2 章由江万寿和呙维撰写,第 7 章由张靖撰写,第 8 章由呙维撰写,其他章节由江万寿撰写。全书由江万寿统稿。

本书力求以应用为导向,使读者在了解开放式遥感软件平台 OpenRS 设计思想的同时,能够掌握该平台提供的基础功能,为平台的应用开发打下基础。

OpenRS 是在"863"计划支持下开发的,没有"863"计划对地观测领域的支持,就不可能有 OpenRS 平台,更不可能有本书的出现,作者谨向国家科学技术部国家遥感中心表示感谢。

同时,作为对课题的配套,武汉大学测绘遥感信息工程国家重点实验室利用其"自主研究经费"设立了"摄影测量与遥感综合处理平台"自主研究项目,使课题组能够更好地凝聚实验室的力量,使该平台得到更好的推广和应用。

从平台的设计开始到本书的完成,作者得到了实验室领导及许多师生的支持、帮助与鼓励,在此表示诚挚的感谢。特别感谢的有:李德仁院士、龚健雅主任、张良培教授、陈晓玲教授、巫兆聪教授、朱庆教授、张过教授、眭海刚教授、刘华副教授、陈玉敏副教授、孙开敏博士、田礼乔博士、李志江博士、刘俊怡博士、刘斌博士、姚璜博士。

同时感谢我的研究助理杨成城、阎吉星,学生王慧贤、颜源、姜三、熊彪等对我的支持和帮助。

限于作者水平,书中难免存在不当之处,欢迎读者批评指正。

江万寿

2015 年 12 月

目 录

《信息科学技术学术著作丛书》序
前言
第1章　绪论 ··· 1
　1.1　遥感数据处理软件的需求 ··· 1
　　1.1.1　遥感数据处理的需求 ··· 1
　　1.1.2　遥感数据处理领域的角色与关系 ·· 2
　1.2　开放与开源 ·· 3
　　1.2.1　开源只是手段，开放才是目的 ·· 3
　　1.2.2　开放系统的OCP ·· 4
　　1.2.3　开放系统的典型代表——Android ··· 5
　1.3　OpenRS设计目标 ·· 6
第2章　OpenRS的总体架构设计与实现思路 ·· 8
　2.1　设计思路 ··· 8
　　2.1.1　功能分层架构 ·· 8
　　2.1.2　插件层次架构 ·· 9
　　2.1.3　桌面处理、分布式处理、网络服务一体化 ································ 11
　2.2　实现思路 ·· 11
　　2.2.1　面向对象设计 ··· 11
　　2.2.2　基于网络服务的分布式处理 ·· 15
　　2.2.3　基于处理链的影像处理 ··· 16
　2.3　OpenRS对象体系基础 ··· 17
　　2.3.1　OpenRS的树状对象体系 ··· 17
　　2.3.2　OpenRS的接口与对象的命名约定 ··· 22
　　2.3.3　对象的查询 ·· 22
　2.4　OpenRS对象的生命周期 ·· 23
　　2.4.1　对象的创建 ·· 23
　　2.4.2　对象的持有与释放 ··· 24
第3章　插件系统设计与实现 ··· 27
　3.1　插件系统的设计与实现 ··· 27
　　3.1.1　OpenRS通用插件体系结构 ·· 28

3.1.2　插件扫描与对象注册过程 …………………………… 28
　　3.1.3　插件对象的查询与创建 …………………………… 32
3.2　平台无关的通用插件系统基础 ………………………………… 34
　　3.2.1　日志服务 …………………………………………… 34
　　3.2.2　XML 序列化服务 …………………………………… 36
　　3.2.3　错误服务(lastErrorService) ………………………… 37
3.3　OpenRS 的三种插件 …………………………………………… 37
　　3.3.1　算法插件 …………………………………………… 37
　　3.3.2　界面扩展插件 ……………………………………… 37
　　3.3.3　属性控件插件 ……………………………………… 38
3.4　OpenRS 的其他基础服务与对象 ……………………………… 39
　　3.4.1　RDF 服务 …………………………………………… 39
　　3.4.2　系统便利服务 ……………………………………… 41
　　3.4.3　矩阵与向量模板类 ………………………………… 44
　　3.4.4　可链接对象 ………………………………………… 54
　　3.4.5　可执行对象接口 …………………………………… 55
3.5　插件开发初步 …………………………………………………… 56
　　3.5.1　插件对象编写 ……………………………………… 56
　　3.5.2　对象注册 …………………………………………… 58
　　3.5.3　对象的创建 ………………………………………… 62

第 4 章　面向遥感影像处理的基础模块 ……………………………… 65

4.1　影像处理模块 orsImage ………………………………………… 65
　　4.1.1　影像源接口 orsIImageSource ……………………… 65
　　4.1.2　影像处理链接口 orsIImageChain …………………… 67
　　4.1.3　影像服务接口 orsIImageService …………………… 68
　　4.1.4　写影像接口 ………………………………………… 71
　　4.1.5　影像数据的读取与处理 …………………………… 72
4.2　影像几何处理模块 orsImageGeometry ………………………… 75
　　4.2.1　遥感应用中影像几何处理的需求 ………………… 76
　　4.2.2　成像几何模型的统一表示 ………………………… 76
　　4.2.3　遥感影像的基本几何处理 ………………………… 77
　　4.2.4　影像几何模型接口 ………………………………… 77
　　4.2.5　多个影像光束的交会 ……………………………… 80
　　4.2.6　空间参考与坐标变换 orsSRS ……………………… 81
　　4.2.7　动态影像几何变换——imageSourceWarper ………… 85

4.3 影像元数据处理模块 …… 90
4.3.1 不同传感器影像的元数据 …… 91
4.3.2 元数据接口设计 …… 91
4.3.3 传感器 …… 92
4.3.4 观测平台 …… 95
4.3.5 影像元数据 …… 96
4.3.6 元数据的读取 …… 98
4.3.7 太阳天顶角、方位角 …… 98
4.3.8 观测天顶角、方位角 …… 99
4.4 简单要素矢量模块 orsSF …… 99
4.4.1 简单要素矢量数据源 …… 100
4.4.2 简单要素矢量层 …… 101
4.4.3 简单要素服务 …… 102
4.5 其他模块 …… 105
4.5.1 基础地理数据管理模块 orsGeoData …… 105
4.5.2 几何变换处理模块 orsGeometry …… 108

第5章 界面扩展设计 …… 111
5.1 BCGControlBar 简介 …… 111
5.1.1 选择 BCGControlBar 的理由 …… 112
5.1.2 BCGControlBar 的扩展性 …… 112
5.2 OpenRS 界面扩展接口与实现 …… 114
5.2.1 框架扩展接口 orsIGuiExtension …… 115
5.2.2 抽象框架接口 orsIFrameWnd …… 116
5.2.3 OpenRS 界面元素 …… 117
5.2.4 框架扩展实现模板 orsIFrameWndHelper …… 129
5.2.5 消息处理与 ID 和谐 …… 134
5.2.6 插件中的界面对象创建与消息处理 …… 142
5.3 属性界面 …… 145
5.3.1 BCG 属性 …… 145
5.3.2 OpenRS 自定义属性界面 …… 149
5.3.3 OpenRS 属性事件的响应与动态属性界面 …… 157
5.4 语言本地化 …… 164
5.4.1 MFC 的本地化方法 …… 164
5.4.2 OpenRS 的本地化方案 …… 165
5.4.3 OpenRS 的本地化的实现 …… 165

第 6 章　桌面集成环境设计与实现 ········· 166
6.1　OpenRS 主控模块 ········· 166
6.2　对象执行器——orsExeRunner ········· 168
6.3　基于图层的影像、矢量显示 ········· 169
6.3.1　基本显示架构 ········· 169
6.3.2　多视图显示的需求分析 ········· 171
6.3.3　多图层、多视图快速显示架构 ········· 172
6.3.4　影像图层及渲染 ········· 186
6.3.5　矢量图层及渲染 ········· 192
6.4　综合显示与集成环境——orsViewer ········· 194
6.4.1　设计目标 ········· 194
6.4.2　设计思路与界面设计 ········· 194
6.4.3　orsViewer 扩展点设计 ········· 195
6.4.4　orsViewer 扩展插件示例 ········· 196

第 7 章　处理流程设计与实现 ········· 197
7.1　外存型处理流程——可执行对象处理流 ········· 197
7.1.1　可执行对象 ········· 197
7.1.2　实现机制 ········· 198
7.2　基于构件模型的可视化流程定制 ········· 199
7.2.1　概述 ········· 199
7.2.2　节点设计 ········· 200
7.2.3　节点关系 ········· 200
7.2.4　工作流的序列化 ········· 201
7.2.5　工作流的执行 ········· 201
7.2.6　OpenRS 中的工作流插件 ········· 202
7.3　面向处理流程的分布式批处理 ········· 202
7.3.1　原理 ········· 202
7.3.2　实现 ········· 203
7.3.3　执行 ········· 204

第 8 章　分布式并行处理环境设计与实现 ········· 205
8.1　PTR 并行模型与功能特性 ········· 205
8.1.1　MapReduce ········· 205
8.1.2　功能特性 ········· 207
8.2　PTR 并行处理框架 ········· 207
8.2.1　系统角色组成 ········· 207

| 8.2.2　并行调度原理 …………………………………………………… 210
| 8.2.3　并行算法提交 …………………………………………………… 215
| 8.2.4　执行状态监控 …………………………………………………… 219
| 8.3　PTR 并行编程接口模型 …………………………………………………… 223
| 8.3.1　接口定义 ………………………………………………………… 223
| 8.3.2　数据交换与信息输出 …………………………………………… 225
| 8.3.3　并行算法插件实例 ……………………………………………… 227
| 8.4　基于 ROI 属性的自动并行机制 …………………………………………… 235
| 8.4.1　遥感数据并行处理的特点 ……………………………………… 235
| 8.4.2　简单任务与并行细分任务的统一 ……………………………… 238
| 8.4.3　基于 ROI 的自动并行化属性定义 ……………………………… 239

第 9 章　网络服务包装与嵌入应用 ……………………………………………… 241
| 9.1　一键式网络服务包装 ……………………………………………………… 241
| 9.1.1　基本思想 ………………………………………………………… 241
| 9.1.2　具体实现 ………………………………………………………… 241
| 9.1.3　一键自动包装 …………………………………………………… 243
| 9.1.4　实现效果 ………………………………………………………… 244
| 9.2　嵌入应用技术 ……………………………………………………………… 245
| 9.2.1　场景对象接口 …………………………………………………… 245
| 9.2.2　场景对象实例 …………………………………………………… 246

第 10 章　插件开发实践 ………………………………………………………… 250
| 10.1　插件开发的粒度 …………………………………………………………… 250
| 10.2　粗粒度插件——可执行对象 ……………………………………………… 250
| 10.2.1　orsISimpleExe 与 orsIParallelExe ……………………………… 250
| 10.2.2　对象命名规则建议 ……………………………………………… 251
| 10.2.3　可执行对象帮助模板 …………………………………………… 252
| 10.2.4　实例——中值滤波 ……………………………………………… 258
| 10.3　细粒度插件（依赖于 OpenRS 遥感处理对象体系）……………………… 274
| 10.3.1　算法对象 ………………………………………………………… 274
| 10.3.2　影像链节点——影像源对象 …………………………………… 274
| 10.3.3　从算法到可执行对象 …………………………………………… 274
| 10.3.4　界面扩展——GuiExtension ……………………………………… 289
| 10.4　一个最小的分布式处理算法软件与客户端构成 ………………………… 289
| 10.5　不同公司或部门软件集成部署方式 ……………………………………… 290

参考文献 …………………………………………………………………… 292
附录 A　OpenRS 宏定义与模板 …………………………………………… 294
　A.1　接口定义宏 ……………………………………………………… 294
　A.2　对象实现宏 ……………………………………………………… 294
　　A.2.1　无主接口的定义和实现 …………………………………… 294
　　A.2.2　带有主接口的定义和实现 ………………………………… 295
　A.3　插件注册宏 ……………………………………………………… 296
附录 B　OpenRS 常用模板 ………………………………………………… 298
　B.1　ref_ptr 的定义 …………………………………………………… 298
　B.2　影像链节点对象帮助模板 ……………………………………… 300
附录 C　OpenRS 编译环境与运行环境 …………………………………… 316
　C.1　OpenRS 编译环境 ……………………………………………… 316
　　C.1.1　目录结构 …………………………………………………… 316
　　C.1.2　第三方库目录 ……………………………………………… 317
　C.2　OpenRS 运行环境 ……………………………………………… 319
　　C.2.1　桌面运行环境 ……………………………………………… 319
　　C.2.2　分布式处理配置 …………………………………………… 321
　　C.2.3　服务网关的安装于配置 …………………………………… 322

第1章 绪 论

进入21世纪以来,遥感对地观测技术已经能够对全球进行多层次、多视角、多领域的观测,遥感技术在国土、测绘、环境、国防等领域的作用也越发凸显。随着遥感信息获取能力的不断提升,新的应用需求也不断涌现,传统的遥感应用技术正发生着革命性的转变,具体体现在以下三个方面:首先,数据获取平台由单一平台向星载、机载、车载与地面等多种平台组成的空天地传感网转变;其次,数据处理模式由单一卫星或航空遥感影像的处理转变为多源、多尺度数据的同化与融合处理;再次,数据处理模式由传统的单一影像处理向整个地区乃至全球多源数据的高性能实时处理系统转变。

针对这种转变,国际上遥感商用软件的发展已经开始发生变化,例如,法国的像素工厂(Pixel Factory)的出现显著地改变了传统摄影测量的数据处理方式,使摄影测量的数据处理水平和处理性能出现了革命性的变化,在国内外产生了巨大的影响(邢诚等,2008)。在遥感数据处理方面,德国Definiens的影像分析软件从传统桌面版的eCognition发展到基于网络服务架构的企业级Definiens EII,实现了在客户端和服务器端上的影像自动或半自动智能信息提取。国内在遥感数据处理服务平台技术上的研究相对滞后,与国外先进水平还存在一定差距。

随着我国遥感对地观测系统建设步伐不断加快,遥感系列、环境系列、资源系列等卫星的成功发射,传统的遥感数据处理软件在当前遥感数据日益多样化、处理流程日益复杂化、应用需求专业化的现状面前,已凸显出诸多不足,迫切需要开展新型的遥感数据处理服务平台架构研究,以满足当前数据来源多样、处理过程复杂、应用需求多变,以及处理数据量大、研究人员分散等需求,实现海量多源遥感数据的按需快速处理,提升我国多源遥感数据综合处理能力与应用服务能力。

在国家"863"计划重点项目的支持下,武汉大学测绘遥感信息工程国家重点实验室研制了开放式遥感数据处理服务平台OpenRS,在平台的功能扩展、性能提升、应用服务等方面取得了技术突破。

1.1 遥感数据处理软件的需求

1.1.1 遥感数据处理的需求

遥感数据处理是一个数据来源复杂、应用目的多样、处理流程多变的专业数据处理过程(陈镜许,2011)。

在数据来源方面,遥感数据处理软件必须能够处理从简单的字节型数据到复杂的复数型数据、从简单的栅格型数据到复杂的矢量型数据、从中心投影的框幅式影像与多中心投影的线阵推扫影像到点扫描的 SAR 影像等不同形式几何模型的影像数据、从人眼可见的光学数据与高光谱数据到人眼不可见的红外数据、从被动的光学数据到主动的雷达数据和激光扫描数据(孙步阳,2009),同时支持不断扩展的影像数据格式。

在处理算法方面,遥感软件平台必须能够支持从简单像元级处理到复杂的对象级处理,包括了传统的影像变换、波段运算、影像滤波、影像融合、影像聚类、影像分类到影像的分割、对象描述、对象分类,从传统的定性处理到逐渐成熟的定量处理等适合不同应用的算法和处理过程。

在处理流程方面,需要面向不同的应用、针对不同的数据来源、选择合适的算法和参数、定制不同的处理流程,形成适用的处理过程和工序。

在处理效率方面,随着遥感数据向高空间分辨率、高光谱分辨率、高时间分辨率的发展,多源遥感数据量越来越大,对遥感数据处理的性能提出了挑战;同时减灾、防灾等应急响应的时间紧迫性,对遥感应用的处理效率提出了小时级的需求。

综上所述,未来的遥感数据处理软件平台必须能够具有高度的扩展性,以支持数据源和处理算法的扩展;必须具有高度的定制性,以满足不同应用目的和数据处理流程构建的需要;必须具有高度的伸缩性,以实现海量数据的快速处理,满足应急响应对海量数据高性能处理的需求。

1.1.2 遥感数据处理领域的角色与关系

遥感处理领域的相关人员一般可分为五类角色,分别是数据提供者、平台架构者、算法开发者、应用整合者和最终用户(刘异等,2009)。数据提供者提供原始数据源;平台架构者进行平台架构设计,并提供给算法开发者和应用整合者调用;算法开发者进行专业算法的开发;应用整合者进行界面、功能与流程的整合;最终用户调用平台功能进行数据处理。传统的遥感数据处理平台软件在开发时,并没有区分这五类角色,平台架构者、算法开发者和应用整合者往往均是同一组织的人员,系统从架构上并没有充分考虑扩展性。为了开发具有更强扩展性、灵活性和分布性的平台,需要合理区分五类角色的行为。数据提供者侧重于提供原始数据;平台架构者侧重于提供基础的遥感数据读写、通用基础处理算法和通用界面元素等;算法开发者侧重于实现专业算法逻辑,实现过程中可利用平台提供的通用处理算法;应用整合者在算法提供者提供的算法和平台架构者提供的基本界面元素基础上,实现各种定制化的服务界面;最终用户在界面开发者提供的界面上,通过算法提供者的算法来处理数据提供者所提供的数据,形成最终所需要的数据结果。图1-1阐述了五类角色之间的相互关系,平台架构者处于整个遥感应用领域的核心地位,通过平台提供的各种功能将其他四类角色有效联系起来。因此,平台设计的优劣,对整个遥感数据处理软件的实现与应用具有重要影响(Guo et al.,2010)。

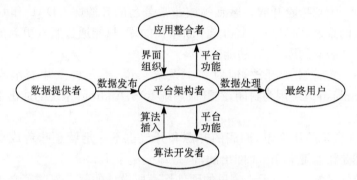

图1-1 遥感数据处理领域角色关系

1.2 开放与开源

由商用遥感软件的发展可以看出,"开放"与"集成"已经成为遥感软件发展的趋势,"伸缩性""扩展性""定制性"已经成为新一代遥感数据处理软件必备的特性。

说到开放,不能不谈到开源。一般认为"开源"是开放的代表,如Linux操作系统、Eclipse开发平台、Android手机操作系统都是非常成功的开源系统。即便在遥感领域,也有OSSIM、Opticks等开源软件。这些软件都以各种方式实现了系统的开放。

但"闭源"是否就意味着封闭呢?应当不是,可以说Windows是最大的开放系统,可以通过挂钩、组件等各种方式,为Windows添加各种各样的功能。同时,苹果公司的IOS是一个既封闭又开放的代表。封闭意味着只能在苹果公司的硬件平台上运行,开放则体现在苹果公司创造的苹果应用商店,世界各地的开发者可以为苹果公司iPhone、iPad开发大量应用。

人们不禁要问,开放和开源到底是什么的关系?开源是否就意味着开放呢?怎样才能实现真正的开放?对于这方面的问题,高焕堂先生做了很好的解答。

1.2.1 开源只是手段,开放才是目的

高焕堂先生认为,"开源只是手段,开放才是目的"(高焕堂,2010)。其主要的观点如下:

(1) 开源只是开放的必要条件,但并非充分条件。
(2) 不一定要开源,但一定要开放。
(3) 开源而不开放,则无法带来商业利益。
(4) 开放不一定要开源,也能带来商业利益。

开源不一定意味着开放。遥感领域非常著名的开源库 GDAL 和 OGR 并不是一个开放的系统,而只是一个功能不断增长的库,只能通过版本更新的方式,通过重新链接新版本的库,实现功能的扩展。

开源不一定意味着开放,但对一个系统来说,不开放意味着消亡。例如,Nokia 的 Symbian 系统因为封闭,在开放系统 IOS、Android 系统的冲击下,已经逐步走向消亡。

同时,在竞争的压力下,彻底的开源和开放也不一定就意味着成功。例如,Meego 虽然是诺基亚和 Intel 推出的,但是却交给了 Linux 基金会来主导,Linux 基金会主导就必然会要求无论哪里的硬件厂商所设计的驱动程序都必须开源,这也正是 Android 跟 Meego 的最大区别。和 Meego 相比,Android 内核本身是开源,但允许开发人员在硬件驱动上保密,企业可以在这一层开发个性化的应用或者形成自身独特的优势,然后再通过这些独特的、个性化的设计获取市场。

因此,从整体来看,必须建立开放的系统,通过开源的方式培育更多的用户,同时能通过闭源的方式为开发者带来一定的利益,才能形成真正的生态环境,在激烈的竞争中生存。

1.2.2 开放系统的 OCP

OCP 是 open closed principle 的缩写,即开放封闭原则。"开放"是指对"扩展"开放,"封闭"是指对"变化"或"修改"封闭(张立新等,2014)。

对"扩展"开放,意味着有新的需求或变化时,可以对现有代码进行扩展,以适应新的情况。

对"变化"或"修改"封闭,意味着类或接口一旦设计完成,就可以独立完成其工作,而不能对类进行任何修改,以保持兼容性。

开放是目的,封闭是必要条件。通过封闭或冻结"接口",才能屏蔽具体的实现细节,满足和实现真正的开放性。一个典型的例子就是 USB 接口的设计及其应用的发展。不管外部设备怎么发展,USB 接口的硬件形式和接口协议是不变的,也就是封闭的。也正是接口协议的封闭,使 U 盘、移动硬盘、刻录机、键盘、鼠标等无数即插即用 USB 设备不断推陈出新(图 1-2)。

OCP 的难点在于识别需要封闭的部分,同时满足扩展和变化的要求。一般需要通过迭代的修改才能逐步形成稳定的接口。

图 1-2　USB 接口及其形状和功能各异的外部设备

1.2.3　开放系统的典型代表——Android

Android 系统就是一个遵守 OCP 原则的典型系统。高焕堂先生把 Android 系统比喻为一棵根深叶茂的大树(图 1-3)。把中间的应用框架比作树干,把丰富多彩的应用比作繁茂的枝叶,把硬件厂商的底层类库和驱动比作吸取硬件营养的深根。

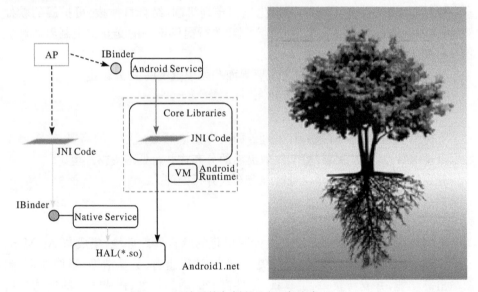

图 1-3　开放系统与树的比拟(高焕堂,2010)

在这里,包括开放式应用的框架、封闭的硬件驱动和应用。通过应用框架屏蔽了底层的硬件驱动和顶层应用之间的关系,确保底层系统(服务、驱动和硬件)具有变动的自由度,吸引更多的厂家加入 Android 阵营;反之,顶层的应用开发者可以不管底层的变化,开发出能够在各家硬件上运行的通用程序。

Android 系统正是通过使底层硬件厂商和顶层应用开发者的利益最大化,而实现了快速的发展。

1.3 OpenRS 设计目标

OpenRS 的设计目标是研制开放式的、具有高度可伸缩性、高度可扩展性、高度可定制性和算法跨平台的通用多源遥感软件公共处理平台,以支持算法、数据源、数据格式、数据类型的动态扩展,支持处理性能的动态扩展、处理流程的动态定制。

1. 可扩展

采用可插拔的软件扩展机制,形成具有"软件总线"功能的平台架构,设计数据源、数据模型、处理算法的接口,形成可替换的算法模块和可扩展的界面模块。

2. 可定制

设计面向遥感数据处理的可视化流程定制机制,结合可插拔、可扩展的软件架构,为不同的遥感应用的处理流程定制和执行提供统一的交互式定制界面和流程执行环境。

支持公用属性配置界面、自主配置界面和处理范围的 ROI 设定。

3. 可伸缩

支持单机处理和分布式并行处理。面向海量遥感数据处理的需要,针对遥感数据处理的特点,形成简单易用、高效灵活的分布式并行处理支持环境。

开发的算法直接支持单机处理和分布式并行处理。

4. 支持图层化的处理

面向遥感数据的显示和处理,设计图层化的综合处理环境,集成与 ArcMap、ERDAS、ENVI 等软件类似的栅格、矢量显示、波段组合、数据渲染、影像卷帘、制图输出等功能,并支持菜单、工具条、控件窗口的动态扩展,构成强大的遥感数据处理基本环境,在此基础上动态集成专业级遥感数据处理模块,形成高效的遥感数据通用处理模块。

5. 易用性

支持桌面调用和处理服务调用。

6. 跨平台

严格区分处理层和界面层,实现处理算法在 Windows 和 Linux 等常用操作系统上编译和运行。

第 2 章 OpenRS 的总体架构设计与实现思路

2.1 设 计 思 路

2.1.1 功能分层架构

现有的通用遥感图像处理平台大多是单机多模块集成式的,所有功能必须运行在同一台机器上,无法发挥多机协同处理的优势。同时,不同的遥感图像数据格式和算法可能互不兼容,难以共享。因此,对遥感图像处理平台这种数据密集、计算密集的处理工具只有建立新的体系结构,才能适应更加大型、复杂和流程化的遥感图像作业环境。OpenRS 拟在遥感数据处理平台的集中/分布式体系、开放式处理结构与可插拔算法服务和协同应用等方面进行突破(吴玮,2014)。

OpenRS 的总体技术路线是基于插件技术和网络服务技术建立开放式、支持分布式并行处理的遥感数据处理软件平台。如图 2-1 所示,OpenRS 软件平台可以粗略地分为基础支撑层、处理插件层和系统应用层。

基础支撑层提供与应用无关的插件管理、并行任务管理与执行、流程定制与执行等基础功能。这部分基础功能是与遥感应用无关的通用功能。而数据管理和显示则与遥感影像有关。为了简化,作为遥感数据处理平台,我们把数据管理与显示也作为基础支撑层的一部分。

处理插件层是与应用有关的业务处理功能实现层。该层强调算法的扩展性和替换性。通过插件系统的支持,可以通过插件机制动态地增加系统的处理功能和处理能力。

系统应用层是直接面向用户的应用系统,包括用户界面和应用逻辑。可以通过流程定制管理器组合处理插件提供的处理功能,形成面向某一具体应用的完整处理链。也可以直接在程序中按编码的形式固化处理链条。在任务的执行上,应用系统可以在本地调用处理节点的功能,也可以把处理任务提交给分布式任务管理器,由分布式任务管理器负责把任务分解分派到集群服务器的处理节点,完成任务的并行处理。

OpenRS 平台架构提供开放式的平台开发环境和分布式处理的架构,但处理算法的质量还需要算法本身和外部数据的质量来保证。例如,正射影像的精度依赖于传感器模型的严密性、传感器参数的准确性、数字地面模型的精度等多个因素。因此,在建立好软件的体系架构后,需要把重心转移到处理算法和具体的业务处理上,以实际数据的生产为推动,发现存在的问题,优化处理方案,提高处理的质量。

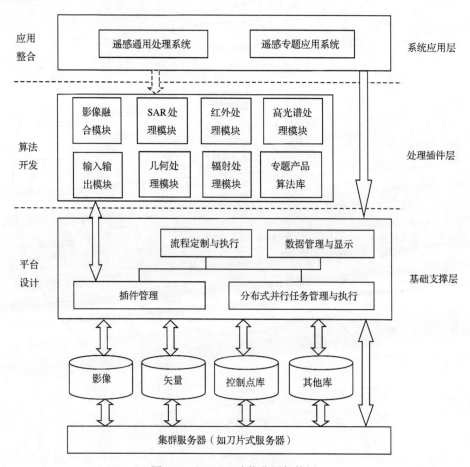

图 2-1 OpenRS 功能分层架构图

2.1.2 插件层次架构

开放式遥感平台采用全插件的系统结构。二次开发人员和平台开发人员一样可以在算法级(包括数据源、几何模型等)、执行模块级、界面级等多个层次对系统功能进行动态扩展(张谦等,2010)。

系统采用类似 Eclipse 的微内核结构,分为插件系统层、服务插件层、对象插件层、应用程序层四个层次(蒋波涛,2008),如图 2-2 所示。

(1)插件系统层:插件管理系统为微内核,提供注册服务、日志服务、错误处理

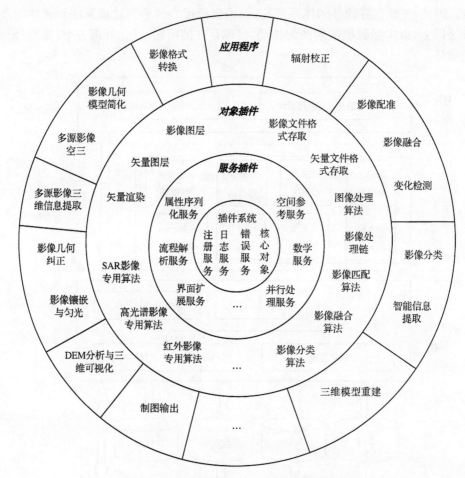

图 2-2 基于全插件架构的遥感软件层次结构

服务三大基础服务。这里的服务概念是指在整个 OpenRS 运行生命周期内唯一的对象,可用于创建其他对象或其他信息的输出(这里的"服务"和 SOA 无关)。插件系统与具体应用无关,相当于"软件总线",是一个通用的架构。

(2) 服务插件层:提供可替代的系统服务或领域相关的模块级服务功能。可替代的系统服务有属性的序列化、流程的解析、界面的扩展、分布式并行处理等。遥感数据处理相关的服务有空间参考、影像处理、矢量处理、图层管理等基础服务。通过服务插件层,系统从一个通用的插件系统扩展服务于某一具体应用领域的基本架构,具备了搭建具体应用程序的基础。服务层的对象是一个单例对象,无需应用程序创建。

(3) 对象插件层:提供可创建的多例对象,用于实现文件读取、传感器几何模型、影像变换、影像分割、影像聚类、影像分类等各种算法对象和其他对象。

(4) 应用程序层:应用程序构建于内核层、服务层、对象层基础之上,通过流程定义、算法配置,结合影像显示等用户界面实现完整的遥感数据处理功能。

2.1.3 桌面处理、分布式处理、网络服务一体化

为了简化软件的复杂性,减少开发的工作量,把代码的一次编写、编译、链接、多处运行作为一个重要的设计目标,使算法的桌面处理代码、分布式并行处理代码、网络服务处理代码完全相同,或者说实际上就是一份复制。

在实现上,遥感数据处理的网络服务调用与本地桌面系统调用分布式数据处理一致。如图 2-3 所示,网络服务网关把远程服务调用转发给作业管理系统,然后作业管理系统充分利用多机处理能力,为远程用户提供处理服务。

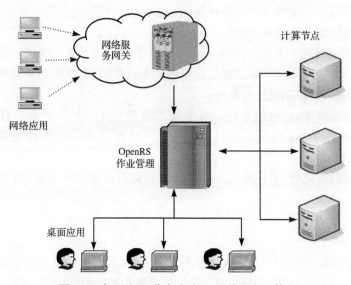

图 2-3 桌面处理、分布式处理、网络服务一体化

服务网关屏蔽了处理服务的实现机制和内部硬件结构,可以随着处理节点的增多不断提高服务能力。

2.2 实现思路

2.2.1 面向对象设计

OpenRS 的基本设计原则是实现遥感数据处理平台与算法的松耦合,达到"插件不和平台链接、插件不和插件链接"的设计要求。主要原因在于框架是易变的,平台框架始终都将经历不断发展演化的过程,需要逐步得到完善。框架是业务

流,可复用性相对更低。而插件是功能模块,不能让模块为平台框架的变化买单。插件模块设计时应忽略框架的存在。

从软件设计的角度看,目前面向对象设计已经发展为组件技术。面向对象技术的基础是封装,即接口与实现分离;面向对象的核心是多态,这是接口和实现分离的更高级升华,使得在运行时可以动态地根据条件来选择隐藏在接口后面的实现。面向对象的表现形式是类和继承(李曙歌,2006),面向对象的主要目标是使系统对象化。对象化的结果就是系统的各部分更加清晰,耦合度大大降低。

面向组件技术建立在对象技术之上,它是对象技术的进一步发展,类概念仍然是组件技术中一个基础概念,但是组件技术更核心的概念是接口。组件技术的主要目标是粗粒度复用,即组件的复用,如一个 dll、一个中间件,甚至一个框架。一个组件可以由一个类或多个类及其他元素组成。但是组件有个很明显的特征,即它是一个独立的物理单元,经常以非源码的形式(如二进制、IL)存在。一个完整的组件中一般有一个主类,而其他的类和元素都是为了支持该主类的功能实现而存在的。为了支持这种物理独立性和粗粒度的复用,组件需要更高级的概念支撑,其中最基本的就是属性和事件。

在 OpenRS 中,组件技术进一步发展为插件技术。插件就是一种特殊的组件。考虑到平台无关性,OpenRS 的插件是一种和 Windows 组件无关的特殊动态库。

总之,接口、属性、事件是 OpenRS 面向对象设计的三大基石,下面分别进行阐述。

1. 面向接口编程

面向接口编程和面向对象编程并不是平级的,它并不是面向对象编程更先进的一种独立的编程思想,它是面向对象编程体系中的思想精髓之一(张祥,2008)。

1) 接口的本质

接口,在表面上是由几个没有主体代码的方法定义组成的集合体,有唯一的名称,可以被类或其他接口所实现(或者也可以说继承)。它在形式上可表示如下:

```
interface InterfaceName
{
    void Method1();
    void Method2(int para1);
    void Method3(string para2,string para3);
}
```

那么，接口的本质是什么，或者说接口存在的意义是什么？可以从以下两个视角考虑：

(1) 接口是一组规则的集合，它规定了实现本接口的类或接口必须拥有的一组规则。体现了自然界"若你是……则必须能……"的理念。

(2) 接口是在一定粒度视图上同类事物的抽象表示。这里强调在一定粒度视图上，是因为"同类事物"这个概念是相对的，它因粒度视图不同而不同。

2) 面向接口编程的目的

面向接口编程强调的是对一个对象的操作只能通过接口进行。为实现同一接口的不同对象，提供了同一类功能的不同实现方法。面向接口编程内在的核心是只要接口不变，对象内部的实现变化就不会影响利用该对象的软件，体现了较高层次的松耦合。

2. 面向属性编程

如果说接口提供了独立于对象具体实现的方法，强调对象不变性，那么属性则在保持对象使用方法稳定性的情况下，提供了一种针对具体对象的动态参数传递方法。目前，C♯等现代编程语言都提供了语言级的属性提取和设置方法。

1) C♯和 Java 的反射与属性

在 C♯中，属性结合了字段和方法的多个方面。对于对象的用户，属性显示为字段，访问该属性需要完全相同的语法。对于类的实现者，属性是一个或两个代码块，表示一个 get 访问器和/或一个 set 访问器。当读取属性时，执行 get 访问器的代码块；当向属性分配一个新值时，执行 set 访问器的代码块。不具有 set 访问器的属性被视为只读属性。不具有 get 访问器的属性被视为只写属性。同时具有这两个访问器的属性是读写属性。

与字段不同，属性不作为变量来分类。因此，不能将属性作为 ref(C♯参考)参数或 out(C♯参考)参数传递。

在 C♯中，反射提供了封装程序集、模块和类型的对象(Type 类型)。可以使用反射动态创建类型的实例，将类型绑定到现有对象，或从现有对象获取类型并调用其方法或访问其字段和属性。如果代码中使用了属性，那么可以利用反射对它们进行访问。

在 Java 运行环境中，对于任意一个类，可以通过 Java 的反射机制动态地获取类的信息。

(1) 在运行时判断任意一个对象所属的类。

(2) 在运行时构造任意一个类的对象。

(3) 在运行时判断任意一个类所具有的成员变量和方法(通过反射甚至可以调用 private 方法)。

(4) 在运行时调用任意一个对象的方法。

C♯的属性通过在内部定义get/set方法,使外部使用时像是在使用变量字段,但其实是在调用get/set方法,以达到透明的封装数据的目的;Java没有属性的概念。Java通过约定为字段XX添加getXX,setXX方法达到同样的目的。

2) OpenRS的对象属性

一般来说,对于标准C++而言是不存在成员属性这个概念的,以前大家都是用getXXX/setXXX来访问或取得数据,好像也没有感觉任何不便。但是用过C♯之类的语言之后,就总觉得C++这个方式太过时了,于是想去实现"属性"这个C++语言缺乏的要素。网络上有很多人已经做了这部分工作,实现的方法有很多种,一种是用模板,一种是根据特定语言来写,如Microsoft Visual C++。但是它们要么很复杂,要么很难记住其准确用法。

如图2-4所示,OpenRS的对象属性其实是一个可动态扩展的多级属性列表(属性树),可通过接口获取、设置或更新,具体实现参见2.3.1节。

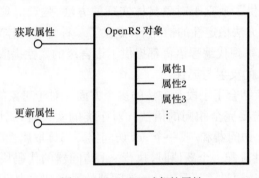

图 2-4　OpenRS 对象的属性

3. 事件处理的委托模型

如图2-5所示,在Windows的组件模型中,经常使用一种称为source-sink的事件处理策略。

(1) COM组件运行时产生一个窗口,当用户双击该窗口时,需要通知调用者。

(2) COM组件用线程方式下载网络上的一个文件,当完成任务后,需要通知调用者。

(3) COM组件完成一个钟表的功能,当预定时间到达时,需要通知调用者。

COM中使用回调接口(包装好的回调函数集)来实现事件的通知。回调接口也称为接收器接口。

使用sink接收器接口的问题是容易造成接口泛滥。若一个客户端需要处理不同来源的事件,则需要继承许多不同来源的接口。使用接口的原因是标准的

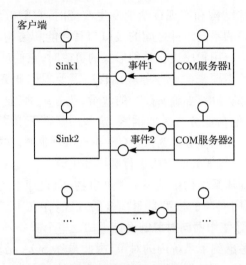

图 2-5 组件的事件接收器接口(sink)

C++没有真正的面向对象的函数指针,导致无法直接调用成员函数指针。

目前,C♯使"委托"的概念日趋流行,简化了松耦合对象的设计模式。在C♯中,委托是一个类,它定义了成员函数(方法)的类型,使得可以将类的成员函数(方法)当做另一个函数(方法)的参数来进行传递,这种将成员函数(方法)动态地赋给参数的做法,可以避免继承大量的 sink 接口,同时使得程序具有更好的可扩展性。同样地,在 Delphi(Object Pascal)中,面向对象的函数指针称为闭包(closure),是 Borland 可视化组建库(visual component library,VCL)的基础。

虽然 C++中没有在编译级定义"委托"类型,但可以通过各种标准的 C++技术或编程技巧实现委托模型。其中最著名的就是 FastDelegate(Clugston,2005)。OpenRS 的委托模型就是直接采用 FastDelegate 来实现事件的处理。

2.2.2 基于网络服务的分布式处理

如果说基于接口的面向对象设计实现的是对象之间的松耦合,那么面向服务的体系结构实现的则是计算机之间的松耦合。

面向服务的体系结构(service-oriented architecture,SOA)是一个组件模型,它将应用程序的不同功能单元(称为服务)通过这些服务之间定义良好的接口和契约联系起来(张晓,2007)。接口是采用中立的方式进行定义的,它独立于实现服务的硬件平台、操作系统和编程语言。这使得构建在各种这样系统中的服务可以一种统一和通用的方式进行交互。

这种具有中立的接口定义(没有强制绑定到特定的实现上)的特征称为服务之间的松耦合。松耦合系统的好处有两点:①它的灵活性;②当组成整个应用程序的每个服务的内部结构和实现逐渐地发生改变时,它能够继续存在。反之,紧耦合意味着应用程序的不同组件之间的接口与其功能和结构是紧密相连的,因而当需要对部分或整个应用程序进行某种形式的更改时,它们就显得非常脆弱。

对松耦合系统的需要来源于业务应用程序需要根据业务的需要变得更加灵活,以适应不断变化的环境,如经常改变的政策、业务级别、业务重点、合作伙伴关系、行业地位以及其他与业务有关的因素,这些因素甚至会影响业务的性质。能够灵活地适应环境变化的业务称为按需(on demand)业务,在按需业务中,一旦需要,就可以对完成或执行任务的方式进行必要的更改。

虽然面向服务的体系结构不是一个新鲜事物,但它却是更传统的面向对象的模型的替代模型,面向对象的模型是紧耦合的,已经存在了二十多年。虽然基于 SOA 的系统并不排除使用面向对象的设计来构建单个服务,但是其整体设计是面向服务的。由于它考虑到了系统内的对象,因此虽然 SOA 是基于对象的,但是作为一个整体,它却不是面向对象的。不同之处在于接口本身。SOA 系统原型的一个典型例子是通用对象请求代理体系结构(common object request broker architecture,CORBA),它已经出现了很长时间,其定义的概念与 SOA 相似。

然而,现在的 SOA 已经有所不同了,因为它依赖于一些更新的进展,这些进展是以可扩展标记语言(extensible markup language,XML)为基础的。通过使用基于 XML 的语言[称为 Web 服务描述语言(Web services definition language,WSDL)]来描述接口,服务已经转到更动态且更灵活的接口系统中,非以前 CORBA 中的接口描述语言(interface definition language,IDL)(董彦卿,2012)可比了。

OpenRS 的分布式处理环境建立在网络服务的基础上,通过 XML 实现分布式并行处理信息的传递和任务调度。

2.2.3 基于处理链的影像处理

影像链的思路是影像链中的对象具有相同的接口,这样对于影像处理算法和显示程序只需要针对"影像源"接口编程,而不需要考虑输入的"影像源"是原始的影像数据块还是处理算法动态生成的影像数据块。影像链模式具有如下优点:

(1) 多个处理过程可以动态链接。

(2) 多个处理过程可以任意组合。

如图 2-6 所示,影像块的读取、变换、滤波、分割、分类和输出等一系列过程可以在线组合。影像显示可以插入影像链的任意位置,用于查看图像处理的当前状态。

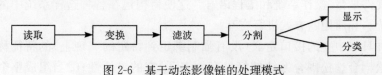

图 2-6 基于动态影像链的处理模式

2.3 OpenRS 对象体系基础

2.3.1 OpenRS 的树状对象体系

1. 根接口

orsIObject 是 OpenRS 的根接口,定义了基本的接口 ID、接口描述、接口查询函数和接口名集合获取等函数。和一般的接口定义不同,getID 和 getDesc 不是纯粹的接口,而是可以被重载的虚函数。OpenRS 的接口形成了一颗对象分类树,getID 和 getDesc 是用以描述每一级接口的。

```
interface orsIObject
{
    ...
    //接口ID定义
    virtual ors_stringgetID()const {return _T("ors");}
    //得到 detailed description
    virtual ors_stringgetDesc()const {return _T("empty");}
    //得到实现的接口类名
    virtual orsArray〈ors_string〉getInterfaceNames()const = 0;
    //根据接口名取得类指针
    virtual void * queryInterface(const orsChar * className)const = 0;
    ...
};
```

2. 对象属性

属性是 OpenRS 中的重要概念。面向接口编程和属性编程是 OpenRS 的实现基础。接口确定了各种对象的分类模型,属性则弥补了接口的不足,用于表示各种动态变化的参数,实现各类具体对象的特定参数描述及其动态定制。orsIProperty定义了可以存取的属性类别。由于属性本身也是一种类别,因此属性对应了 XML 的树状结构。

```
interface orsIProperty : public orsIObject
{
public://单接口
    virtual void getAttr(const orsChar * name,ors_int16 & value)const = 0;
```

```cpp
    virtual bool getAttr(const orsChar * name, ors_int32 &value)const = 0;
    virtual bool getAttr(const orsChar * name, ors_float64 &value)const = 0;

    //传回拷贝
    virtual bool getAttr(const orsChar * name, ors_string &value)const = 0;
    virtual bool getAttr(const orsChar * name, orsVectorBase &value)const = 0;
    virtual bool getAttr(const orsChar * name, orsMatrixBase &value)const = 0;

    //传回内部指针
    virtual bool getAttr(const orsChar * name, const ors_byte * &pValue, ors_int32
        &nLength)const = 0;
    virtual bool getAttr(const orsChar * name, ref_ptr <orsIProperty> &value)const = 0;

public://复接口
    virtual bool getAttr(const orsChar * name, orsArray <ors_int16> &values)const = 0;
    virtual bool getAttr(const orsChar * name, orsArray <ors_int32> &values)const = 0;
    virtual bool getAttr(const orsChar * name, orsArray <ors_float64> &values)const = 0;
    virtual bool getAttr(const orsChar * name, orsArray <ors_string> &values)const = 0;

    //传回内部指针
    virtual bool getAttr(const orsChar * name, orsArray <const ors_byte * > &values,
        orsArray <ors_int32> &vLength)const = 0;
    virtual bool getAttr(const orsChar * name, orsArray <ref_ptr <orsIProperty> >
        &values)const = 0;

public://变接口
    virtual bool setAttr(const orsChar * name, ors_int16 value, ors_uint32 index = 0)= 0;
    virtual bool setAttr(const orsChar * name, ors_int32 value, ors_uint32 index = 0)= 0;
    virtual bool setAttr(const orsChar * name, ors_float64 value, ors_uint32 index =
        0)= 0;
    virtual bool setAttr(const orsChar * name, const orsChar * value, ors_uint32
        index = 0)= 0;
    virtual bool setAttr(const orsChar * name, const ors_byte * pValue, ors_int32
        nLength, ors_uint32 index = 0)= 0;
    virtual bool setAttr(const orsChar * name, const orsVectorBase &value)= 0;
    virtual bool setAttr(const orsChar * name, const orsMatrixBase &value)= 0;
    virtual bool setAttr(const orsChar * name, ref_ptr <orsIProperty> value, ors_
```

```
        uint32 index = 0)= 0;

public://增接口
    virtual void addAttr(const orsChar * name,ors_int16 value)= 0;
    virtual void addAttr(const orsChar * name,ors_int32 value)= 0;
    virtual void addAttr(const orsChar * name,ors_float64 value)= 0;

    //内部复制
    virtual void addAttr(const orsChar * name,const orsChar * value)= 0;
    virtual void addAttr(const orsChar * name,const ors_byte * pValue, ors_int32
        nLength)= 0;
    virtual void addAttr(const orsChar * name,const orsVectorBase &value)= 0;
    virtual void addAttr(const orsChar * name,const orsMatrixBase &value)= 0;

    //内部引用
    virtual void addAttr(const orsChar * name,ref_ptr〈orsIProperty〉value)= 0;

public:
    //属性变换通知回调,为便免循环引用
    virtual void setListener(orsIPropertyListener * listener)= 0;

    //轮询接口,返回//属性个数
    virtual ors_uint32 size()const = 0;

    //属性名称和类型,用于序列化
    //ind是属性的索引号,范围为 0 ~ (size()-1);
    virtual void getAttributeInfo(ors_uint32 ind,ors_string & name, orsVariantType
        & type,ors_int32 &numOfValues)const = 0;

    //接口名(ID)
    ORS_INTERFACE_DEF(orsIObject, _T("property"))
};
```

3. 基于接口关系的层次化的对象 ID

根据遥感影像处理的需要,OpenRS 中设计了影像数据源、影像链处理链、影像采样算法、影像滤波算法、空间参考与坐标变换、影像几何模型、简单要素模型、通用图层模型等公共接口。图 2-7 列出了部分接口及其之间的继承关系。这些接口及其派生的对象具有严格的层次关系。同一接口的对象功能相似,能够在一定程度上相互替代。基于这种层次关系,应用程序可以利用平台核心基础模块查询每个对象实现的接口、对象的名字、对象的描述、对象所在的插件等信息。

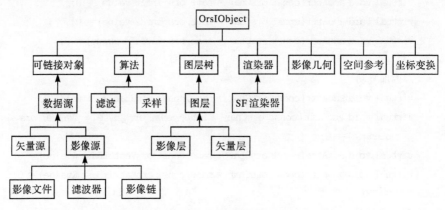

图 2-7 遥感软件公共对象接口体系(部分)

在 OpenRS 中,为了把某一接口的具体实现对象作为可配置的参数,必须能够根据整个唯一的 ID(UUID)创建该对象。正如对象接口体系构成了一棵树一样。如果给每个接口定义一个名字,那么很容易根据这些接口实现一个从对象根接口到对象叶子的分级对象 ID。如图 2-8 所示,orsGDALImageReader 继承了一系列接口,每一级接口都有自己的 ID,则 orsGDALImageReader 的完整 ID 为"ors. dataSource. image. reader. gdal"。这样每一类对象 ID 的前缀是相同的,区分只在最后一级的实现类 ID。当然这样设计的缺点是有可能出现相同的对象 ID,可以通过在实现类 ID 中加上作者和版本签名来解决。

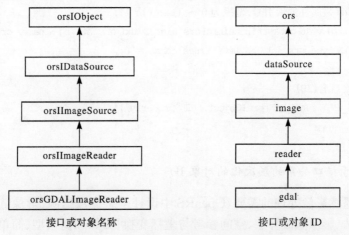

图 2-8 接口 ID 与对象 ID 的构成关系

在 OpenRS 中，接口 ID 的定义通过 ORS_INTERFACE_DEF 宏实现，对象 ID 的定义通过 ORS_OBJECT_IMP1、ORS_OBJECT_IMP2、ORS_OBJECT_IMP3 等宏实现。ORS_OBJECT_IMPn 中 n 代表要能够被查询到的接口个数。

```cpp
interface orsIDataSource : public orsIConnectableObject
{
    //接口 ID 定义
    ORS_INTERFACE_DEF(orsIConnectableObject, _T("dataSource"));
};

interface orsIImageSource : public orsIDataSource
{
    ...
    //接口 ID 定义
    ORS_INTERFACE_DEF(orsIDataSource, _T("image"))
};

class orsIImageSourceReader : public orsIImageSource
{
    ...
    //接口 ID 定义
    ORS_INTERFACE_DEF(orsIImageSource, _T("reader"))
};

Class orsGDALImageReader : public orsIConnectableObjectNoInputHelper <orsIImageS-
    ourceReader>, public orsObjectBase
{
public:
    orsGDALImageReader();
    virtual ~orsGDALImageReader();

    //得到 detailed description
    virtual ors_string getDesc()const {return "基于 GDAL 的影像读取";}

    //实现对象的虚表和 ID 定义
    ORS_OBJECT_IMP3(CGDALImageReader, orsIImageSourceReader, orsIImageSource,
        orsIConnectableObject, "gdal", "GDAL Image Reader")
```

```
    //上面的宏展开后,等价于
    ORS_OBJECT_DEF_NORMAL(orsIImageSourceReader, "gdal", "GDAL Image Reader")
    ORS_BEGIN_VTABLE_MAP(CGDALImageReader)
        ORS_INTERFACE_ENTRY(orsIObject)
        ORS_INTERFACE_ENTRY(orsIImageSourceReader)
        ORS_INTERFACE_ENTRY(orsIImageSource)
        ORS_INTERFACE_ENTRY(orsIConnectableObject)
    ORS_END_VTABLE_MAP
};
```

2.3.2 OpenRS 的接口与对象的命名约定

在一般的设计中,接口类和实现分别采用 XXXXInterface 和 XXXXImplementation 的方式。这种命名方法看似很清晰,但若再加上 namespace 的名字,则容易出现名词过于冗长的情况。另外,过多使用 namespace 也很容易出现名字冲突。因此,在 OpenRS 中,对象命名没有采用 namespace,同时 interface 和 implementation 也采用最简短的缩写。

OpenRS 建议接口类命名以"orsI"开头,接口实现类命名以"orsX"开头。其中,"ors"是"OpenRS"的缩写,"I"代表"Interface","X"代表"Implementation"。

OpenRS 命名的原则是"简约而不简单"。

2.3.3 对象的查询

通过 orsIPlatform 的 getRegisterService 得到 orsIRegisterService 服务,该服务提供查询相关功能。其中,通过 getObjectDescsByInterface 可以按接口名得到所有实现某一接口的实现类列表。而由 getObjectDescByID 可以通过 ID 得到对象描述。

```
interface orsIRegisterService : public orsIService
{
    ...
    //得到兼容某个接口的算法描述
    virtual orsArray〈ref_ptr〈orsIObjectDesc〉〉 getObjectDescsByInterface(const
        orsChar * interfaceName)= 0;

    //通过对象名称得到描述
    virtual ref_ptr〈orsIObjectDesc〉 getObjectDescByID(const orsChar * objID)= 0;
    ...
```

```
};
```
实例:影像读取类查询。
```
orsFileFormatList orsXImageService::getSupportedImageFormats()
{
    orsFileFormatList allFormatlist;

    orsIRegisterService * registerService = getPlatform()-> getRegisterService
        ();

    //查询实现 orsIImageSourceReader 的对象
    orsArray<ref_ptr<orsIObjectDesc>> objDescs =
        registerService-> getObjectDescsByInterface("orsIImageSourceReader");
    ...
}
```

2.4 OpenRS 对象的生命周期

2.4.1 对象的创建

orsIPlatform 的 createObject 提供通过某一对象实现类的 ID 创建该对象的功能。为了减少类型的转换,可以直接通过 OpenRS 提供的 ORS_CREATE_OBJECT 宏来完成对象的创建。

```
ORS_CREATE_OBJECT(classType, objID)
```

实例:orsIImageSourceReader 对象的创建与支持格式的查询。

```
orsFileFormatList orsXImageService::getSupportedImageFormats()
{
    orsFileFormatList allFormatlist;

    //注册服务获取
    orsIRegisterService * registerService = getPlatform()-> getRegisterService();

    //实现 orsIImageSourceReader 接口的对象查询
    orsArray<ref_ptr<orsIObjectDesc>> objDescs =
        registerService-> getObjectDescsByInterface("orsIImageSourceReader");
```

```
for(unsigned i=0;i<objDescs.size();i++)
{
    orsFileFormatList formatlist;
    ref_ptr<orsIObjectDesc> desc = objDescs[i];

    //通过对象 ID 创建该对象
    ref_ptr<orsIImageSourceReader> reader = ORS_CREATE_OBJECT(orsIImage-
        SourceReader,getPlatform(),desc->getID().c_str());

    //创建成功?
    if(reader != NULL){
        //得到格式列表
        formatlist = reader->getSupportedImageFormats();
        for(unsigned j=0;j<formatlist.size();j++)
            allFormatlist.push_back(formatlist[j]);
    }
}
return allFormatlist;
}
```

2.4.2 对象的持有与释放

OpenRS 遵循"内存谁分配,谁释放"的原则。这是为了避免分配和释放内存不是由相同的堆管理程序完成的情况。例如,动态链接库中的堆在默认情况下是由 msvcrt.dll 中的堆管理程序管理的(以动态链接的方式),而 exe 程序的堆在默认情况下由程序自己的代码管理(以静态链接的方式),由于它们的堆管理程序不同,因此当动态链接库分配的内存在 exe 程序中释放时就有可能出错,因为 exe 程序所在的堆并没有分配这块内存,而用户却要求它释放这块内存。

为了实现"内存谁分配,谁释放"的原则,OpenRS 采用具有引用计数的智能指针 referenced pointer(ref_ptr)来实现对象的管理。ref_ptr 克服了 auto_ptr 不能放入 stl 容器的缺点。orsIObject 定义了 addRef()和 release()这两个接口函数来实现智能指针的引用计数。

```
interface orsIObject
{
    ...
```

```
virtual void addRef()= 0;
virtual void release()= 0;
...
};
```

在对象实现时,设计了 orsObjectBase 来辅助实现 addRef()和 release()。例如,对象类 orsHelloObject 继承了 orsObjectBase 来实现接口 addRef()和 release()。

```
class orsHelloObject : public orsIHelloWorld, public orsObjectBase
{
    ...
}
```

orsObjectBase 的定义如下:

```
//用于辅助引用计数实现
class orsObjectBase
{
public:
    orsObjectBase(){m_refCount = 0;}
    virtual ~orsObjectBase(){};
    virtual void internalAddRef(){
        m_refCount++;
    }
    virtual void internalRelease(){
        m_refCount--;
        if(!m_refCount)
            delete this;
    }
protected:
    mutable ors_int32 m_refCount;
};
```

OpenRS 的智能指针是由模板类 ref_ptr 来实现的。在离开 ref_ptr 的定义域后,引用的对象会自动减 1。如果引用计数减到 0,那么对象将自动释放。因此,在 OpenRS 中,对象的使用只需要按 ref_ptr〈〉的方法引用,就可以自动完成智能对象的持有和释放。例如

```
ref_ptr<orsIImageSourceReader> reader = getImageService()-> openImage(...);
```

为了提高调用效率,不需要在所有的地方采用智能指针。可以把智能指针转化为普通指针使用。方法是调用智能指针对象的 get()函数,如 read.get()可以获得智能指针对象 reader 管理的实际指针。智能指针的使用原则是,在会影响生命周期的地方都应该使用智能指针。

第 3 章 插件系统设计与实现

3.1 插件系统的设计与实现

插件(plug-in,又称 addin、add-in、addon 或 add-on,又译外挂)也称为扩展,是一种遵循一定规范的应用程序接口编写出来的程序,主要用来扩展软件功能,很多软件都有插件,有些由软件公司自己开发,有些则是第三方或软件用户个人开发。

插件的本质在于不修改程序主体(平台)的情况下对软件功能进行扩展与加强,当插件的接口公开后,任何公司或个人都可以制作自己的插件来解决一些操作上的不便或增加新的功能(蒋波涛,2008;吴亮等,2006),也就是实现真正意义上的"即插即用"软件开发。平台+插件软件结构是将一个待开发的目标软件分为两部分:一部分为程序的主体或主框架,可定义为平台;另一部分为功能扩展或补充模块,可定义为插件。

开始时,很容易混淆插件与对象的关系,甚至把对象说成插件。实际上,对象不依赖于插件,插件只是对象的容器。自从面向对象的概念出现以后,对象的概念就已经深入人心,虽然和插件没有必然的关系,但是插件作为对象的容器,可以创造出新的对象,使系统的对象可以在不需要编译连接的情况下,不断增长。通过软件模块的动态插入获得新的功能,充分体现了系统的扩展性。

对应用开发来说,不需要知道对象的位置,甚至不管它来自哪个插件,也不管它是否来自插件。只需要通过已知的对象 ID 或给定的对象 ID,以统一的方式动态创建对象。要说基于插件的开发方式有什么特点,那就是只能针对对象的接口进行编程,而不能对具体的实现类进行编程。对实现类直接编程,意味着对该类的动态链接,这个只能在同一个插件内实现,因此不鼓励使用这种方式。有时编程人员也会偷懒,直接在同一个插件内通过实现类的方式直接创建对象,但总会发现若需要对程序进行重构,如调整实现类到其他插件,就必须对这个调用类进行修改。

插件系统由插件管理器和插件组成。OpenRS 的插件采用 C++语言实现。插件宿主的形式为动态链接库,在 Windows 平台的后缀为.dll,Linux 平台的后缀为.so。插件管理器负责插件的扫描、对象的注册、查询与创建。

3.1.1 OpenRS 通用插件体系结构

如图 3-1 所示,插件系统的核心是插件管理器。对插件来说,插件管理器负责插件的扫描、插件内对象的注册、插件内对象的创建等基本功能;对应用程序来说,插件管理器提供对象的查询和对象的注册。

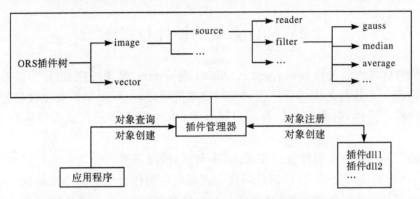

图 3-1 OpenRS 通用插件体系结构

由于插件的扫描、对象的注册和对象的创建是和应用无关的,因此是通用的,且和应用无关。3.1.2 节开始将详细介绍插件系统的设计和实现过程。

3.1.2 插件扫描与对象注册过程

OpenRS 的插件放在 plugin 子目录下,插件管理功能包含在 orsBase 动态库中(图 3-2)。orsBase 动态库包含平台对象 orsXPlatform、插件注册对象 orsXRegisterService。插件扫描和对象注册通过 orsXPlatform 和 orsXRegisterService 来实现。orsXPlatform 管理 orsXRegisterService 等基本服务,orsXRegisterService 负责组织插件动态库、插件中的对象和对象工厂。

在 OpenRS 的应用程序启动时,通过调用 orsInitliaze()初始化插件系统,结束时调用 orsUninitialize 释放装载的插件(图 3-3)。程序执行过程中,通过调用注册服务创建需要的对象、然后执行需要的功能。不同的对象组合调用构成程序的完整功能。

1. 插件的正常加载

插件的扫描和加载由 orsXRegisterService 对象完成。主要包括插件目录扫描、插件动态库加载、插件内含对象注册等过程(图 3-4)。

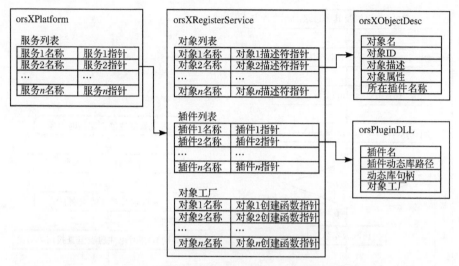

图 3-2　平台对象、注册对象与插件的关系

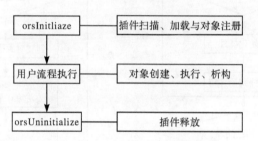

图 3-3　OpenRS 程序基本执行流程

2. 插件系统的序列化

插件注册表为一棵插件树，插件序列化的第一步就是把插件的信息添加到插件树上。OpenRS 的插件树就是一个 OpenRS 的属性树 pluginTree。图 3-5 所示的插件序列化除了把插件名称、ID 等信息加到插件属性节点，还要实现插件内部对象的序列化到插件属性节点。对象的序列号除了把对象名称、ID 等信息加到对象属性节点，还要把对象实现的接口列表加到对象属性节点。

插件系统序列化的第二步就是把插件树写入 XML 文件。插件树写入 XML 文件通过 OpenRS 属性树 XML 序列化器实现。XML 序列化器将在后面讲到。下面为只有一个插件时的插件系统序列化文件。多个插件时的文件结构是一样的，只是有多个 plugin 记录。

〈OpenRS_Plugin_System〉

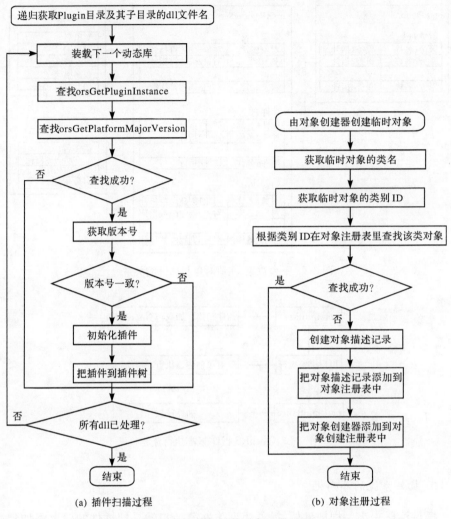

(a) 插件扫描过程　　(b) 对象注册过程

图 3-4　插件扫描与对象注册过程

⟨First_start_up type="string"⟩ true ⟨/First_start_up⟩
⟨Plugin⟩
　　⟨ID type="string"⟩ org.openRS.2DFeatureLayer ⟨/ID⟩
　　⟨Name type="string"⟩ ors2DFeatureLayer.dll ⟨/Name⟩
　　⟨Object⟩
　　　　⟨Desc type="string"⟩ 2D Feature Map Layer ⟨/Desc⟩
　　　　⟨ID type="string"⟩ ors.layer.2DFeature ⟨/ID⟩
　　　　⟨Interface type="string"⟩ orsIObject ⟨/Interface⟩

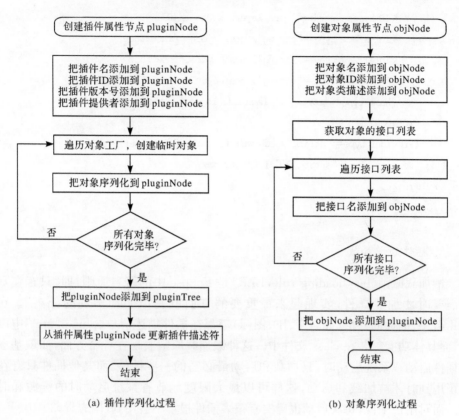

(a) 插件序列化过程　　　　　　　　(b) 对象序列化过程

图 3-5　插件的序列化

　　⟨Interface type="string"⟩ orsILayer ⟨/Interface⟩
　　⟨Interface type="string"⟩ orsILayerTreeItem ⟨/Interface⟩
　　⟨Name type="string"⟩ 2D Feature Layer ⟨/Name⟩
⟨/Object⟩
⟨Object⟩
　　⟨Desc type="string"⟩ GCP Points Map Layer ⟨/Desc⟩
　　⟨ID type="string"⟩ ors.layer.GCP ⟨/ID⟩
　　⟨Interface type="string"⟩ orsIObject ⟨/Interface⟩
　　⟨Interface type="string"⟩ orsILayer ⟨/Interface⟩
　　⟨Interface type="string"⟩ orsILayerTreeItem ⟨/Interface⟩
　　⟨Name type="string"⟩ GCP Points Layer ⟨/Name⟩
⟨/Object⟩
⟨Object⟩
　　⟨Desc type="string"⟩ Tie Points Map Layer ⟨/Desc⟩

```
        <ID type="string">ors.layer.tiePts</ID>
        <Interface type="string">orsIObject</Interface>
        <Interface type="string">orsILayer</Interface>
        <Interface type="string">orsILayerTreeItem</Interface>
        <Name type="string">Tie Points Layer</Name>
    </Object>
        <Provider type="string">edu.whu.liesmars</Provider>
        <Version type="string">0.1</Version>
</Plugin>
</OpenRS_Plugin_System>
```

3. 插件的懒加载

懒加载原则(lazy loading rule,LLR)是 Eclipse 中的著名原则,即"只有在真正需要时才加载插件,实现起来最重要的方面就是声明和实现的分离"。在 Eclipse中,插件的外形(如名字、ID、图标)等都在插件描述清单"plugin.xml"中声明,而具体功能封装在 class 文件中。这种懒加载原则表现在各个方面,如最基本的插件启动。系统启动时,只加载和启动最必需的一些插件,而其他插件只有在真正用到时才被加载和启动,这样可以最大限度地节省系统启动时的资源和时间。而对用户来说,每次启动也确实有很多插件根本不会用到(闫志贵,2010)。

在 OpenRS 中,懒加载的策略比较简单。在启动一个应用程序、调用 orsInitliaze 初始插件系统时,可以设置是否进行完整的插件扫描,进行对象注册。如果注册表已经存在,那么可以不进行插件扫描,而在创建某一个对象时再加载该插件(图 3-6)。

3.1.3 插件对象的查询与创建

1. 根据接口名查询实现的对象

对象的查询,特别是实现了某一接口的对象的查询在开放式软件平台中具有重要作用,是平台扩展的核心功能。

2. 对象的创建

对象创建是插件系统的基本功能,对象创建过程要考虑懒加载模式,对于没有加载的插件要在创建对象时动态加载(图 3-7)。

第 3 章 插件系统设计与实现

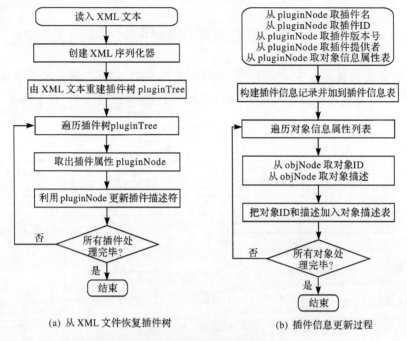

(a) 从 XML 文件恢复插件树　　(b) 插件信息更新过程

图 3-6　懒加载与插件信息更新

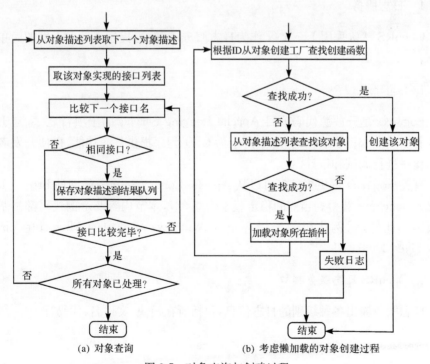

(a) 对象查询　　(b) 考虑懒加载的对象创建过程

图 3-7　对象查询与创建过程

3.2 平台无关的通用插件系统基础

插件系统由日志服务、错误服务、注册服务与 XML 序列化服务四个部分组成（图 3-8）。因为与具体处理无关，所以可以通用于不同的服务。

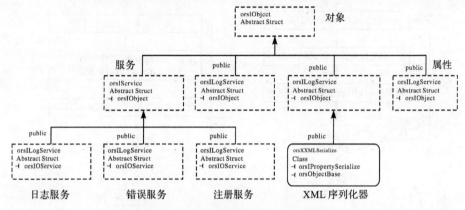

图 3-8 平台无关的通用插件系统组成

3.2.1 日志服务

OpenRS 默认采用 Log4cxx 作为日志服务，同时也提供了针对界面的简单对话框。

1. Log4cxx

Log4cxx 是开放源代码项目 Apache Logging Service 的子项目之一，是 Java 社区著名的 log4j 的 C++ 移植版，用于为 C++ 程序提供日志功能，以便开发者对目标程序进行调试和审计。

有关 Log4cxx 的更多信息可以从 Apache Loggin Service 的网站 http://logging.apache.org 获得。当前的稳定版本为 0.9.7，本节内容及示例代码都是基于此版本。此外，示例代码的编译环境为 Windows 环境中的 Microsoft Visual C++.Net 2003。

2. OpenRS 日志服务接口

日志服务输出各种级别的日志信息，包括所在模块、文件名、代码行。

```
interface orsILogService:public orsIService
{
```

```cpp
public:

    //致命错误信息输出
    virtual void fatal(const orsChar * strModule, const orsChar * msg, const orsChar
        * file = NULL, int row = 0)= 0;

    //错误信息输出
    virtual void error(const orsChar * strModule, const orsChar * msg, const orsChar
        * file = NULL, int row = 0)= 0;

    //警告信息输出
    virtual void warn(const orsChar * strModule, const orsChar * msg, const orsChar *
        file = NULL, int row = 0)= 0;

    //一般信息输出
    virtual void info(const orsChar * strModule, const orsChar * msg, const orsChar *
        file = NULL, int row = 0)= 0;

    //调试信息输出
    virtual void debug(const orsChar * strModule, const orsChar * msg, const orsChar
        * file = NULL, int row = 0)= 0;

    //是否允许交互式消息框
    virtual void enableInteractive(bool bEnable = true)= 0;

    //接口名称
    ORS_INTERFACE_DEF(orsIService, _T("log"))
};
```

3. orsIPlatfrom 的 logPrint

OpenRS 还在 orsIPlatform 中直接提供了一个简略版的日志输出接口 logPrint。该接口能够接收类似 printf 格式的日志信息。

```cpp
virtual int logPrint(orsLogLEVEL loglevel, const orsChar * fmt, ...)= 0;
```

日志级别 orsLogLEVEL 定义为枚举类型。

```cpp
enum orsLogLEVEL{
```

```
    ORS_LOG_DEBUG,
    ORS_LOG_INFO,
    ORS_LOG_WARNING,
    ORS_LOG_ERROR,
    ORS_LOG_FATAL
};
```

3.2.2 XML 序列化服务

OpenRS 注册表的序列化文件格式采用 XML 格式,许多遥感数据的元数据文件也为 XML 格式。因此,XML 序列化服务是 OpenRS 的核心服务之一。OpenRS 的 XML 引擎采用 TinyXML,内置在 orsBase 中。

1. TinyXML

如图 3-9 所示,TiXmlBase 是所有类的基类,TiXmlNode、TiXmlAttribute 两个类都继承自 TiXmlBase 类,其中 TiXmlNode 类指的是所有被〈…〉…〈…/〉包括的内容,而 XML 中的节点又具体分为以下几方面内容,分别是声明、注释、节点及节点间的文本,因此在 TiXmlNode 的基础上又衍生出 TiXmlComment、TiXmlDeclaration、TiXmlDocument、TiXmlElement、TiXmlText、TiXmlUnknown 这几个类,分别用来指明具体是 XML 中的哪一部分。TiXmlAttribute 类不同于 TiXmlNode,它指的是在尖括号里面的内容,如〈… * * * =…〉,其中 * * * 就是一个属性。

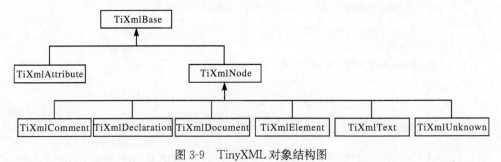

图 3-9 TinyXML 对象结构图

2. 序列化服务 orsXXMLSerialize

orsXXMLSerialize 调用 TinyXML 的功能进行属性树的序列化(生成 XML 字符串)和反序列化(从 XML 字符串生成属性树)。属性类型对照见表 3-1。

表 3-1　OpenRS 属性类型和 XML 类型对照表

OpenRS 属性类型	XML 类型串	备注
ORS_V_UNKOWN		无对应
ORS_V_BOOL	bool	TiXmlText
ORS_V_I2	int16	TiXmlText
ORS_V_I4	int32	TiXmlText
ORS_V_R8	float64	TiXmlText
ORS_V_STR	string	TiXmlText
ORS_V_BLOB	blob	TiXmlText
ORS_V_CHILD		TiXmlElement
ORS_V_VECTOR	vector	TiXmlText
ORS_V_MATRIX	matrix	TiXmlText
ORS_V_OBJECT		无对应，只用于内存
ORS_V_LISTNER		无对应，只用于内存

3.2.3　错误服务（lastErrorService）

设置错误信息，返回最后一次的错误信息。

3.3　OpenRS 的三种插件

开放式遥感软件遥感平台的插件对象分为三种类型：①用于算法或模型的可替换性对象，统称为"算法对象"；②用于功能性扩展的插件，称为"扩展点对象"；③用于控件扩充的插件，称为"属性控件插件"。

3.3.1　算法插件

算法对象，继承于某一个算法接口，可以有很多不同的实现。算法插件形成某个算法的具体实现列表，可以用于流程的定制，供流程定制者选择用哪一种具体的算法实现具体的功能（图 3-10）。

3.3.2　界面扩展插件

"扩展点"是 Eclipse 的概念，主要用于界面的扩展，在应用程序加载时把指定类型的界面扩展插件全部加载（梁冰，2012）。这一类型的扩展也可用于算法插件的参数定制，用于在桌面系统中为插件提供参数定制的界面扩展。而在分布式处理的节点上，和界面相关的插件都不会出现。

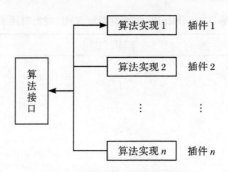

图 3-10　OpenRS 的算法插件

开放式遥感平台的一个重要研究目标是扩展性，软件模块之间的松耦合是保证系统动态扩展的关键。

OpenRS 通过插件技术实现了遥感数据处理平台的高度"松耦合"，把平台和功能模块完全松绑，不进行模块的静态链接（图 3-11）。

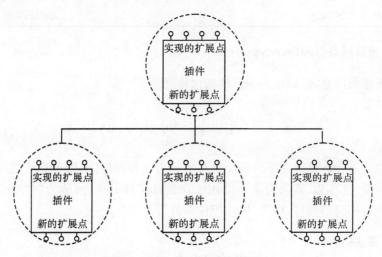

图 3-11　OpenRS 的界面扩展点

3.3.3　属性控件插件

属性控件插件是 OpenRS 特有的一种控件扩充功能，可以在 OpenRS 的参数对话框中灵活地把各种 MFC 的控件作为属性添加上到 BCG 的属性界面中，而不用针对不同属性参数单独编写对话框代码来实现，开发者用起来很方便。如图 3-12 所示，可以利用属性控件插件实现数学形态学中核矩阵的编辑，而这个控件对应于属性名"KernelMatrix"，即只要属性名为"KernelMatrix"，就可以在编辑时弹出这个对话框进行核矩阵的编辑。

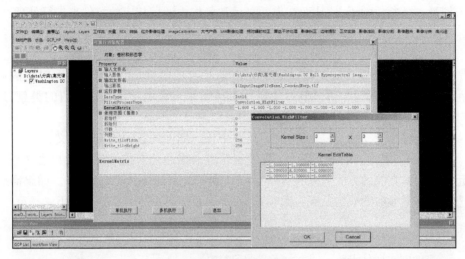

图 3-12　属性控件插件

3.4　OpenRS 的其他基础服务与对象

为了减少模块数,作者把平台无关、应用无关的基础功能放在 OpenRS 基础模块 orsBase 中,该模块编译后直接放在 bin 目录下。除了完成插件目录 module 和 plugins 的动态链接库扫描、对象注册、对象创建等工作,orsBase 中还提供了 RDF 服务、便利服务及向量和矩阵模板等功能。

3.4.1　RDF 服务

RDF 是 W3C 组织推荐的国际标准,用来描述互联网上的任意资源,实现不同平台和操作系统的信息共享(刘建明,2012)。RDF 提供了连接两个数据元素,并用一条术语来描述两者之间关系的能力。

1. 语义与三元组

两个数据元素(资源)和描述术语的组合构成了一个形如主谓宾的三元组(triple)声明。主语和宾语中间的谓语也就是描述连接的术语,也称为三元组的"属性"。例如,OpenRS 中可以引入类似〈ImageFile〉〈isA〉〈File〉这样的声明。其中,"ImageFile"是主语,"File"是宾语,而"isA"则是将以上两者捆绑在一起的谓语。

通过存储 RDF 格式文件,并对其运用一定的规则,就可以推导出新的信息来,表达出这些数据相关的清晰的来龙去脉。在 OpenRS 中,也正是利用这一点来刻画一个参数的语义,从而在界面显示和操作中调用相应的控件来进行处理。

2. OpenRS 资源描述服务

对象属性以属性树的形式表达了对象拥有的属性,可以用于提取对象的属性,并设置对象属性的值。但对象属性只描述了对象属性的名称和值的类型,这对于对象参数值的约束和对象链接的约束判断远远不够。

一种解决途径是,在属性定义或描述中指定属性的各种约束和语义。

OpenRS 基于本体论的思路,通过三元组来描述公有或私有的参数语义。三元组以文本文件的方式放置在 RDF 目录\OpenRS\desktop\etc\RDF。如果需要把采样类型参数约束为从下拉菜单选择,那么可以再用三元组描述如下:

⟨SampleMethod⟩⟨IsA⟩⟨EnumString⟩
⟨SampleMethod⟩⟨Desc⟩⟨Sample Method String⟩
⟨SampleMethod⟩⟨Enum⟩⟨Nearest⟩
⟨SampleMethod⟩⟨Enum⟩⟨Bilinear⟩
⟨SampleMethod⟩⟨Enum⟩⟨BiCubic⟩

在 OpenRS 中,RDF 三元组在内存中以属性表的方式进行存储,形成一个属性树。以上述 SampleMethod 三元组为例,在内存中相关的三元组被自然地组织为一棵树,同类的属性如 Enum 被合并为一个列表(图 3-13)。因此,在内存中 RDF 描述的所有三元组就是一棵属性树。

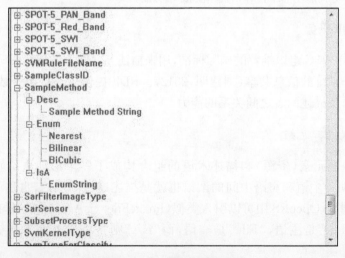

图 3-13　OpenRS 三元组列表

基于 RDF 属性树，OpenRS 资源描述服务提供三元组查询、语义推理查询等功能。接口如下：

```
//资源描述服务
interface orsIRDFService : public orsIService
{
public:
    //返回主体的属性
    virtual orsIProperty * getSubject(const orsChar * subjectName) const = 0;

    //按推理模式查询主体的属性
    virtual bool getObject(const orsChar * subject, const orsChar * verb, orsString
        &object) const = 0;
    //查询满足条件的父类
    virtual bool getParentSubject(const orsChar * subject, const orsChar * verb,
        const orsChar * objectWanted, orsString &parent) const = 0;
    //从 XML 文件中读取元数据，并存到属性树中
    virtual ref_ptr〈orsIProperty〉readMetaDataFromXmlFile(const orsChar * xml-
        FileName)= 0;

public:
    ORS_INTERFACE_DEF(orsIService, _T("rdf"))
};
```

其中，getSubject 是最基本的操作，在实现上，该方法等同于从 RDF 属性树查找一个子属性。getSubject(subjectName)返回以 subjectName 为主词的相关三元组。例如，getRdfService()-> getSubject("SampleMethod")返回 SampleMethod 为根节点的属性树。

getObject(const orsChar * subject, const orsChar * verb, orsString &object)
getParentSubject(subject, verb, objectWanted, &parent)根据

另外，readMetaDataFromXmlFile 用于把 XML 文件读到内存中，形成一棵属性树。

3.4.2 系统便利服务

系统便利服务 orsIUtilityService 用于获取 OpenRS 目录、操作系统的目录相关操作、文件相关操作。其具体包括以下功能。

1) OpenRS 目录获取功能

（1）GetDirectory_EXE：OpenRS 执行程序所在目录，即 orsBase 所在目录。

（2）GetDirectory_ETC：OpenRS 的 etc 目录。etc 是 OpenRS 中用于存放空间参考、语言本地化、RDF 服务等数据的目录。

2) 目录和文件操作

（1）CheckFileExist：检测文件是否存在。

（2）CheckFolderExist：检测目录是否存在。

（3）CopyFile：把一个文件复制为另一个文件。

（4）DeleteFile：删除一个文件。

（5）CopyFolder：复制一个目录为另外一个目录。

（6）DeleteFolder：删除一个目录。

（7）MoveFile：把一个文件移动到另一个位置。

（8）MoveFolder：把一个目录移动到另一个位置。

3) 获取系统内存

GetAvailableMemory：获取可分配的系统内存。

4) 执行外部程序

ShellExecute：执行外部程序。

5) 宏替换

（1）AddMarco：增加一个宏。

（2）GetMarco：获取一个宏。

（3）ReplaceMacroString：用输入属性中的文件名替换输出属性中的宏。

6) 数字字符串转换

（1）String2Integers：从"1,2,3"，"1-3"等数字字符串中提取数字，转换为数字列表。

（2）Integers2String：整数数组转换为"1,2,3"，"1-3"形式的字符串表示。

```
//便利服务接口

interface orsIUtilityService : public orsIService
{
public:
    virtual bool GetDirectory_EXE(orsString &dir)= 0;
    virtual bool GetDirectory_ETC(orsString &dir)= 0;


    virtual bool ShellExecute(const orsChar *cmdLine, bool bWaite = true)= 0;

    //用 inputProp 中字符串替换 outputProp 中的宏
```

```cpp
    virtual bool ReplaceMacroString(orsIProperty * outputProp, orsIProperty * in-
        putProp, orsIProperty * paraProp = NULL)= 0;

    virtual void AddMarco(const orsChar * macroName, const orsChar * macroValue)
        = 0;
    virtual bool GetMarco(const orsChar * macroName, orsString & (macroValue)= 0;

    //从"1,2,3", "1-3" 等数字字符串中提取数字
    virtual orsArray<ors_int32> String2Integers(orsString &strIntegers)=0;

    //整数数组转换为"1,2,3", "1-3"形式的字符串表示
    virtual orsString Integers2String(orsArray<ors_int32> &integers)=0;

    virtual ors_int64 GetAvailableMemory()= 0;
    virtual orsDataTYPE GetOutputDataType(ors_string &strOutputDataType)=0;

    virtual bool CheckFileExist(const orsChar * filePath)= 0;
    virtual bool CheckFolderExist(const orsChar * filePath)= 0;

    virtual bool CopyFile(const orsChar * filePath, const orsChar * newFilePath,
        bool bFailIfExists)= 0;
    virtual bool DeleteFile(const orsChar * filePath)= 0;

    virtual bool CopyFolder(const orsChar * srcDir, const orsChar * dstDir, bool
        bFailIfExists)= 0;
    virtual bool DeleteFolder(const orsChar * filePath)= 0;

    //移动文件夹或目录:在同一盘符时可提高效率
    virtual bool MoveFile(const orsChar * filePath, const orsChar * newFilePath,
        bool bFailIfExists)= 0;
    virtual bool MoveFolder(const orsChar * srcDir, const orsChar * dstDir, bool
        bFailIfExists)= 0;

public:
    //接口 ID
    ORS_INTERFACE_DEF(orsIService, _T("utility"))
};
```

3.4.3 矩阵与向量模板类

1. 矩阵模板

头文件包含 #include "orsBase/orsMatrix.h"。

```
template <typename _T>
class orsMatrix : public orsMatrixBase
{
    ...
}
typedef orsMatrix<double> orsMatrixD;
typedef orsMatrix<float>  orsMatrixF;
```

orsMatrix 实现了常用的矩阵功能。以矩阵 mA(4,3)、mB(4,3)、mC(4,3)、mAt(3,4)、mAtA(3,3)为例,说明如下。

1) 矩阵定义

```
orsMatrixD mA(4,3);
orsMatrixD mB(4,3);
orsMatrixD mC(4,3);
orsMatrixD mAt(3,4);
orsMatrixD mAtA(3,3);
```

2) 矩阵转置

矩阵 mA 转置存入 mB,有两种形式:

```
mA.transpose(mAt);
mAt = mA.transpose();
```

3) 矩阵清 0 或填充 1

```
mA.Zero();
mA.Ones();
```

4) 矩阵的迹

```
double trace = mAtA.Trace();
```

5) 求代数余子式

```
orsMatrixD mA11 = mA.AlgbComplement(1, 1);
```

6) 求矩阵的行列式

```
double det = mA.det();
```

7) 是否方阵

```
bool bIs = mA.isSquare();
```

8) 元素乘

两个矩阵对应元素相乘，与 Matlab 点乘相同。

```
orsMatrixD mAB;
mAB = mA.elementMultiple(mB);
```

9) 元素除

两个矩阵对应元素相除，与 Matlab 点除相同。

```
orsMatrixD mAB;
mAB = mA.elementDivide(mB);
```

10) 矩阵相加

```
template <typename _T>
orsMatrix<_T> operator + (const orsMatrix <_T> &mA, const orsMatrix<_T> &mB)
{
    assert(mA.Rows()==mB.Rows()|| mA.Cols()==mB.Cols());

    orsMatrix<_T> mC;
    mC.Alloc(mA.Rows(), mA.Cols());

    for (int row = 0; row < mC.Rows(); row++)
    {
        for (int col = 0; col < mC.Cols(); col++)
        {
            mC[row][col] = mA[row][col] + mB[row][col];
        }
    }

    return mC;
```

};

11) 矩阵相减

```cpp
template <typename _T>
orsMatrix<_T> operator - (const orsMatrix<_T> &mA, const orsMatrix<_T> &mB)
{
    assert(mA.Rows()== mB.Rows()|| mA.Cols()== mB.Cols());

    orsMatrix<_T> mC;
    mC.Alloc(mA.Rows(), mA.Cols());

    for (int row = 0; row < mC.Rows(); row++)
    {
        for (int col = 0; col < mC.Cols(); col++)
        {
            mC[row][col] = mA[row][col] - mB[row][col];
        }
    }

    return mC;
};
```

12) 矩阵相乘

```cpp
template <typename _T>
orsMatrix<_T> operator * (const orsMatrix<_T> &mA, const orsMatrix<_T> &mB)
{
    assert(mA.Cols()== mB.Rows());

    orsMatrix<_T> mC;
    mC.Alloc(mA.Rows(), mB.Cols());

    for (int row = 0; row < mC.Rows(); row++)
    {
        for (int col = 0; col < mC.Cols(); col++)
        {
            mC[row][col] = 0;
            for(int k=0; k < mA.Cols(); k++)
```

```cpp
                mC[row][col] += mA[row][k] * mB[k][col];
            }
        }

        return mC;
    };
```

2. 向量模板

```cpp
class orsVectorBase
{
protected:
    int m_nRows;
    orsDataTYPE m_dataType;
    void * m_pBuf;

protected:
    orsVectorBase(): m_nRows(0), m_dataType(ORS_DT_UnKNOWN), m_pBuf(NULL){};
    virtual ~orsVectorBase()  {if(m_pBuf) delete (char * )m_pBuf;};

    void SetDataType(int bytes)  {
        if(bytes == 4)
            m_dataType = ORS_DT_FLOAT32;
        else
            m_dataType = ORS_DT_FLOAT64;
    };

public:
    void SetDataType(orsDataTYPE dataType){
        m_dataType = dataType;
    }

    orsDataTYPE GetDataType() const{return m_dataType;};
    int Rows() const{return m_nRows;};
    void * Buf() {return m_pBuf;};
    const void * Buf() const{return m_pBuf;};
};
```

```cpp
template <typename _T>
class orsVector : public orsVectorBase
{
public:
    orsVector(){
        SetDataType(sizeof(_T));
    };

    orsVector(int nRows){
        SetDataType(sizeof(_T));
        Alloc(nRows);
    }

    orsVector(_T a, _T b, _T c){
        SetDataType(sizeof(_T));
        Alloc(3);

        _T *buf = (_T *)Buf();

        buf[0] = a; buf[1] = b; buf[2] = c;
    }

    orsVector(_T a, _T b, _T c, _T d){
        SetDataType(sizeof(_T));
        Alloc(4);

        _T *buf = (_T *)Buf();

        buf[0] = a; buf[1] = b; buf[2] = c; buf[3] = d;
    }

    //复制构造函数
    orsVector(const orsVector <_T> &vec){
        SetDataType(sizeof(_T));
        Alloc(vec.m_nRows);
        memcpy(m_pBuf, vec.m_pBuf, m_nRows*sizeof(_T));
    }
```

```cpp
~orsVector()   {DeAlloc();}

void DeAlloc()
{
    if(m_pBuf)
        delete (_T*)m_pBuf;
    m_pBuf = NULL;
}

bool Alloc(int nRows)
{
    if(nRows == m_nRows)
        return true;

    DeAlloc();
    m_pBuf = new _T[nRows];
    if(NULL == m_pBuf)
        return false;
    m_nRows = nRows;

    return true;
}

void Zero()
{
    if(NULL != m_pBuf)
        memset(m_pBuf, 0, m_nRows * sizeof(_T));
}

void Ones()//填充1
{
    if (NULL!=m_pBuf)
    {
        _T * tempBuf= (_T *)m_pBuf;
        for (int i=0;i< m_nRows;i++)
            tempBuf[i]=_T(1.0);
    }
}
```

```
_T & operator [] (int i)
{
    assert(i < m_nRows);
    return ((_T *)m_pBuf)[i];
}

const _T& operator [] (int i)const
{
    assert(i < m_nRows);
    return ((_T *)m_pBuf)[i];
}

//赋值操作符
orsVector& operator= (const orsVector &vec)
{
    //避免自我赋值
    if(this == &vec)
        return *this;

    Alloc(vec.m_nRows);
    memcpy(m_pBuf, vec.m_pBuf, m_nRows * sizeof(_T));

    return *this;
}

//归一化
void Unit()
{
    double s = 0;
    _T *t = (_T *)m_pBuf;

    int i;
    for (i =0; i < m_nRows; i++)
    {
        s += *t **t;t++;
    }

    assert(0 != s);
```

```cpp
        s = sqrt(s);
        t = (_T *)m_pBuf;
        for (i = 0; i < m_nRows; i++)
        {
            *t /= s;t++;
        }
    }

    orsVector unit()
    {
        orsVector vA = *this;

        vA.Unit();

        return vA;
    }

    //数量积
    double dotProduct(orsVector &vB)
    {
        assert(m_nRows == vB.Rows());
        assert(m_dataType == vB.m_dataType);

        _T *a=(_T)m_pBuf;
        _T *b= (_T *)vB.Buf();

        double c = 0;
        for (int row = 0; row < m_nRows(); row++)
        {
            c += *a **b;
            a++;b++;
        }

        return c;
    }

    //叉积,仅限于三维向量
    orsVector cross(orsVector &vec)
```

```
{
    assert(3 == m_nRows && 3 == vec.Rows());
    assert(m_dataType == vec.m_dataType);

    _T *a=(_T *)m_pBuf;
    _T *b= (_T *)vec.Buf();

    orsVector<_T> c(3);
    c[0] = a[1]*b[2] - b[1]*a[2];
    c[1] = -(a[0]*b[2] - b[0]*a[2]);
    c[2] = a[0]*b[1] - b[0]*a[1];

    return c;
}

void CopyData(_T *buf)
{
    memcpy(buf, m_pBuf, m_nRows*sizeof(_T));
}

_T *Buf(){return (_T *)m_pBuf;};

//元素乘,与Matlab点乘相同,对应元素相乘,结果还是向量
orsVector<_T> elementMultiple(const orsVector<_T> &vB)
{
    assert(Rows()!= vB.Rows());

    orsVector<_T> &vA = *this;
    orsVector<_T> vC(Rows());
    for (int row = 0; row < vC.Rows(); row++)
        vC[row] = vA[row] * vB[row];

    return vC;
};

//元素除,与Matlab点除相同
orsVector<_T> elementDivide(const orsVector<_T> &vB)
{
```

```cpp
        assert(Rows()!= vB.Rows());

        orsVector<_T> &vA = *this;
        orsVector<_T> vC(Rows());
        for (int row = 0; row < vC.Rows(); row++)
            vC[row] = vA[row]/vB[row];
    };

};


template <typename _T>
orsVector<_T> operator+ (const orsVector<_T> &vB1, const orsVector<_T> &vB2)
{
    orsVector<_T> vRes(vB1);
    if (vB1.Rows()!= vB2.Rows())
        return vRes;

    for (int row = 0; row < vRes.Rows(); row++)
        vRes[row] = vRes[row] + vB2[row];

    return vRes;
};

template <typename _T>
orsVector<_T> operator- (const orsVector<_T> &vB1, const orsVector<_T> &vB2)
{
    orsVector<_T> vRes(vB1);
    if (vB1.Rows()!= vB2.Rows())
        return vRes;

    for (int row = 0; row < vRes.Rows(); row++)
        vRes[row] = vRes[row] - vB2[row];

    return vRes;
};

template <typename _T>
orsVector<_T> operator* (const orsMatrix<_T> &mA, const orsVector<_T> &vB)
```

```
{
    assert(mA.Cols()== vB.Rows());
    
    orsVector<_T> vC(mA.Rows());
    for (int row = 0; row < mA.Rows(); row++)
    {
        vC[row] = 0;
        for (int col = 0; col < mA.Cols(); col++)
        {
            vC[row] += mA[row][col] * vB[col];
        }
    }
    
    return vC;
};

typedef orsVector<double> orsVectorD;
typedef orsVector<float> orsVectorF;
```

3.4.4 可链接对象

可链接对象接口是一个重要的接口,表示一种输入、输出可以关联的抽象对象类型,如影像内存处理对象可以设计为可链接对象,以实现透明式的软件开发。通过可链接对象接口可以把不同的处理算法在内存中串接起来,形成一个综合的内存处理单元。

```
interface orsIConnectableObject: public orsIObject
{
public:
    //判断当前的链接对象是否能在objectIdx的位置链接上object(如算法对象)
    virtual bool canConnect(orsIConnectableObject * object)= 0;
    
    //链接上输入
    virtual bool connect(orsIConnectableObject * object)= 0;
    
    //在输入中断开一个对象
    virtual bool disconnect(orsIConnectableObject * object)= 0;
```

```
//断开所有输入
virtual void disconnectAll()= 0;

//设置该链接点对象的真正逻辑算法内置对象实现
virtual bool setLogicObject(orsIObject *obj)= 0;

//得到逻辑对象
virtual ref_ptr<orsIObject> getLogicObject()= 0;

//遍历接口
virtual ors_int32 getNumberOfInput()= 0;

//通过索引得到输入链接对象
virtual orsIConnectableObject * getInputObjectByIndex(unsigned idx)= 0;

ORS_INTERFACE_DEF(orsIObject, "connectable")
};
```

3.4.5 可执行对象接口

可执行对象是具有完整输入、输出文件的处理单元,可以通过数据文件的关联形成工作流。可执行对象侧重于外存中的数据处理,而可链接对象侧重于内存中的数据处理。

```
interface orsIExecute : public orsIObject
{
public:
    //输入文件
    virtual orsIProperty * getInputFileNames()= 0;

    //内部参数
    virtual orsIProperty * getParameterArgs()= 0;

    //输出文件
    virtual orsIProperty * getOutputFileNames()= 0;

    virtual bool setArguments(orsIProperty * inputFileNames, orsIProperty * parame-
        terArgs, orsIProperty * outputFileNames)= 0;
```

```
//序列化接口,在 orsIObject 中已定义
/virtual const orsIProperty *getProperty()const = 0;

//输入参数信息
virtual bool initFromProperty(orsIProperty *property)= 0;

//取自定义算法配置界面的对象 ID,默认没有自定义界面
//注意:不在算法内部直接实现该配置界面,应把配置界面作为一个扩展在 GUI 插件
里单独实现,则影响算法插件的移植性
virtual const orsChar *getConfigDlg(){return NULL;};

ORS_INTERFACE_DEF(orsIObject, _T("execute"));
};
```

3.5 插件开发初步

3.5.1 插件对象编写

在 OpenRS 系统中,每个插件对应一个操作系统级别的动态链接库(Windows 上为 *.dll,Linux 上为 *.so),而在每个插件内部包含若干数据源、算法或界面对象(即一个 C++类),每个对象对应一种特定的功能,如图 3-14 所示。

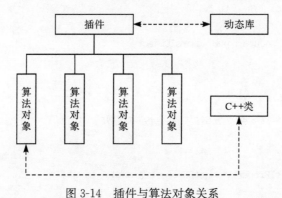

图 3-14　插件与算法对象关系

为了实现算法对象状态的保存与动态调用,每个插件内的对象需要通过属性来实现序列化和反序列化。属性可以看成一个类似 XML 结构的树(事实上,属性 XML 可以序列化为一个 XML 文档),每个节点具有类型,并允许嵌套(即允许子树)(图 3-15)。每个节点代表了算法对象的成员变量(图 3-16)。

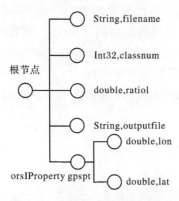

图 3-15 树装组织的对象描述

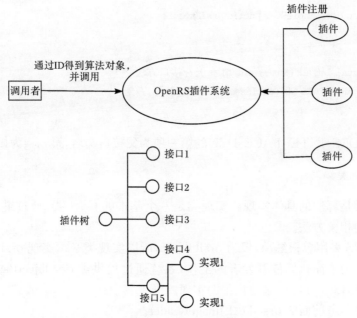

图 3-16 树装组织的接口、对象描述与调用

1. 编写一个插件的步骤

(1) 工程创建：在 VC 下创建一个 DLL 工程（如果是界面扩展插件，那么需要创建 MFC DLL 工程），并设置 C++ 运行时为 multithread dll。

(2) 插件外壳 orsXPlugin.cpp 编写：为了便于插件查询，插件需要提供自身的元信息，包括编写者、版本、插件名称、插件 ID 等，为此，OpenRS 系统提供了一个 orsIPlugin 接口，让插件编写者通过 orsXPlugin 继承来填充元信息。

(3) 算法对象编写:插件外壳编写完毕后,开始编写算法对象本身,算法对象的编写分为接口类编写与实现类编写两个部分。如果该算法对象是实现一个已有的接口,那么只需完成实现类的编写,如滤波。编写接口类主要在某些 OpenRS 目前接口定义不完整而需要补充时才需要。

(4) 在 orsXPlugin 中的 initialize 函数注册算法对象。OpenRS 注册采用创建临时对象的方式实现。

(5) 编译与部署。

2. 接口类编写

接口类为一个普通的 C++ 纯虚类,不同点在于需要继承自 orsIObject 或其派生类,并加上接口定义宏 ORS_INTERFACE_DEF,形式如下。

```
interface orsISimpleExe: public orsIExecute
{
public:
    virtual int add(int a, int b)= 0;//接口方法
    ORS_INTERFACE_DEF(orsIExecute, "simple");//宏
};
```

其中,ORS_INTERFACE_DEF 的第一项为父接口类名,第二项为接口 ID。

3. 实现类编写

实现类是算法的具体实现。实现类是一个基本的 C++ 类,通过继承接口来实现接口的相关方法。

OpenRS 采用单根继承,根为 orsIObject,所以实现类必须实现 orsIObject 的所有方法。为了简化插件开发者的工作,系统通过提供类 orsObjectBase 和系列宏 ORS_OBJECT_IMP*n* 来提供帮助实现。

例子请参考前面的 orsGDALImageReader。

3.5.2 对象注册

首先编写创建函数,然后在 onsXPlugin 对象的 Initialize 函数中,注册创建函数。

```
…
#include "GdalImageReader.h"
//gdal 影像读取对象创建函数
orsIObject *createGdalImageReader(bool bForRegister)
```

```cpp
{
    CGDALImageReader * reader = new CGDALImageReader;
    return reader;
}


#include "GDALImageSourceWriter.h"
//gdal影像源写对象创建函数
orsIObject * createGdalImageSourceWriter(bool bForRegister)
{
    CGDALImageSourceWriter * writer = new CGDALImageSourceWriter;
    return writer;
}


#include "GDALImageWriter.h"
//gdal影像写对象创建函数
orsIObject * createGdalImageWriter(bool bForRegister)
{
    CGDALImageWriter * writer = new CGDALImageWriter;
    return writer;
}


class orsXPlugin: public orsIPlugin
{
public:
    //插件描述
    virtual const orsChar * getID()
    {
        return "org.openRS.imageProcess";
    };

    //插件名
    virtualconst orsChar * getName()
    {
        return "image Process";
    };

    //插件提供者
    virtual const orsChar * getProvider()
```

```
    {
        return "edu.whu.liesmars";
    };

    //插件版本
    virtual const orsChar * getVersion()
    {
        return "0.1";
    }

    //插件初始化
    virtual bool initialize(orsIPlatform * platform)
    {
        orsIRegisterService * pRegister = platform->getRegisterService();

        //对象工厂注册
        pRegister->registerObject(createGdalImageReader);
        pRegister->registerObject(createGdalImageSourceWriter);
        pRegister->registerObject(createGdalImageWriter);
        return true;
    }

    virtual void finalize()
    {
    }
};

ORS_REGISTER_PLUGIN(orsXPlugin)
```

1. 插件的编译与部署

设置插件的链接目录为/openRS/desktop/bin/目录(debug/vc60、release/vc60、debug/vc90、release/vc90)下的 Plugins 目录。

目前,OpenRS 平台还没有实现插件目录的动态监测功能,需要手工删除过期的插件描述文件 plugintree.xml。OpenRS 系统启动时,平台初始化会自动重新扫描所有插件并注册插件。

2. 插件对象的可视化查看

插件和插件对象都登记在 OpenRS 内部注册表中，可以查看已注册的插件。打开 OpenRS 主程序，单击主菜单中的插件\插件管理器，就能得到插件对象浏览对话框。按接口名依次展开为对象树，可以找到插件对象，如图 3-17 和图 3-18 所示。

图 3-17　OpenRs 主程序

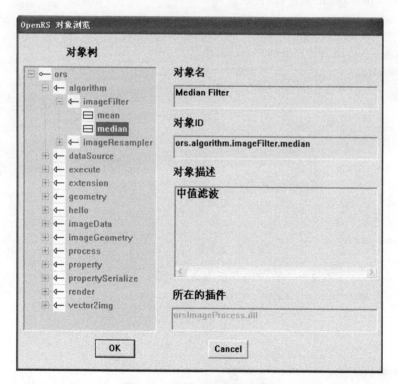

图 3-18　插件对象浏览对话框

这里看到的是中值滤波的算法对象。从对话框中可以看到，中值滤波的对象名为"Median Filter"，对象 ID 为"ors. algorithm. imageFilter. median"，描述为"中值滤波"，所在插件动态库为"orsImageProcess. dll"。

3.5.3 对象的创建

OpenRS 应用程序使用插件对象的方式是通过 orsIPlatfrom 查询、创建对象并调用该对象的功能。任何要使用 OpenRS 插件平台的应用程序必须链接 orsBaseD.dll 或 orsBase.dll。

当然也可以采用动态库自主装载并查询 orsInitialize() 函数的方式。这可以进一步实现应用程序和 orsBase 的松耦合。

1. 平台初始化与终止

调用 OpenRS 的 orsInitialize() 方法对插件平台进行初始化,得到 orsIPlatform 平台接口指针,该指针是插件平台的唯一入口,退出插件平台时,调用 orsUninitialize() 终止平台,以释放资源。

以 imageViewer 为例,说明如下:

```
//保存平台指针,便于利用 getPlatform 调用
orsIPlatform * m_pPlatform = NULL;
orsIPlatform * getPlatform()
{
    return m_pPlatform;
}


CImageViewerApp::CImageViewerApp()
{
    //启动平台
    ors_string errorinfo;

    //第二个参数说明是否重写扫描插件,注册对象
    m_pPlatform = orsInitialize(errorinfo, true);
    …
}


CImageViewerApp::~CImageViewerApp()
{
    …
    //结束平台
    orsUninitialize();
}
```

2. 对象的查询

通过 orsIPlatform 的 getRegisterService 得到 orsIRegisterService 服务,该服务提供查询相关功能。其中,通过 getObjectDescsByInterface 可以按接口名得到所有实现某一接口的实现类列表,而通过 getObjectDescByID 可以通过 ID 得到对象描述。

```
interface orsIRegisterService : public orsIService
{
    ...
    //得到兼容某个接口的算法描述
    virtual orsArray<ref_ptr<orsIObjectDesc>> getObjectDescsByInterface(const
        orsChar * interfaceName)= 0;

    //通过对象名称得到描述
    virtual ref_ptr<orsIObjectDesc> getObjectDescByID(const orsChar * objID)= 0;
    ...
};
```

实例:影像读取类查询。

```
orsFileFormatList orsXImageService::getSupportedImageFormats()
{
    orsFileFormatList allFormatlist;
    orsIRegisterService * registerService = getPlatform()->getRegisterService();

    //查询实现 orsIImageSourceReader 的对象
    orsArray<ref_ptr<orsIObjectDesc>> objDescs =
        registerService->getObjectDescsByInterface("orsIImageSourceReader");
    ...
}
```

3. 对象的创建

orsIPlatform 的 createObject 提供通过某一对象实现类的 ID 创建该对象的功能。

为了减少类型的转换,可以直接通过 openRS 提供的 ORS_CREATE_OBJECT 宏来完成对象的创建。

```
ORS_CREATE_OBJECT(classType, pPlatform, objID)
```

实例：orsIImageSourceReader 对象的创建与支持格式的查询。

```cpp
orsFileFormatList orsXImageService::getSupportedImageFormats()
{
    orsFileFormatList allFormatlist;

    //注册服务获取
    orsIRegisterService *registerService = getPlatform()->getRegisterService();

    //实现 orsIImageSourceReader 接口的对象查询
    orsArray<ref_ptr<orsIObjectDesc>> objDescs =
        registerService->getObjectDescsByInterface("orsIImageSourceReader");

    for(unsigned i=0;i< objDescs.size();i++)
    {
        orsFileFormatList formatlist;
        ref_ptr<orsIObjectDesc> desc = objDescs[i];

        //通过对象 ID 创建该对象
        ref_ptr<orsIImageSourceReader> reader = ORS_CREATE_OBJECT(orsIImageS-
            ourceReader, getPlatform(),desc->getID().c_str());

        //创建成功?
        if(reader != NULL){
            //得到格式列表
            formatlist = reader->getSupportedImageFormats();
            for(unsigned j=0;j< formatlist.size();j++)
                allFormatlist.push_back(formatlist[j]);
        }
    }
    return allFormatlist;
}
```

第 4 章　面向遥感影像处理的基础模块

遥感影像处理最基本的是影像数据本身的处理,首先要解决的是各种格式的影像读取问题;然后在此基础上,编写各种影像数据处理算法;在涉及影像的地理定位时,影像纠正等几何处理操作需要涉及影像的几何模型,实现从影像坐标到地面坐标或地面坐标到影像坐标的变换;如果需要对两个以上的影像数据或其他数据进行处理,那么影像几何模型的空间参考就成为一个不可忽视的问题;在考虑影像的辐射时,影像成像的太阳位置、传感器位置等因素则变得至关重要。因此,作为遥感影像数据处理的基础,本章首先要对遥感影像处理涉及的影像源接口、影像成像几何模型接口、空间参考接口、元数据接口进行较详细的讨论。在此基础上,将给出基本的应用。

4.1　影像处理模块 orsImage

影像处理模块主要提供影像读写服务,接口包括影像源接口、影像处理链接口、影像服务接口和写影像接口。

4.1.1　影像源接口 orsIImageSource

影像源接口定义了一个影像文件源或经过一系列处理的影像源的可能操作。通过该接口能获得一个影像源高度和宽度、影像源的波段数、影像源的数据类型、影像源的几何成像模型接口、影像数据块的读取接口和影像波段的元数据接口。理论上,在元数据方面,应该返回影像元数据接口,但在使用上更加关心和波段本身有关的元数据。特别地,如果需要从影像元数据取波段元数据,那么会涉及波段索引等问题。为了减少复杂性,选择直接返回和数据有关的元数据。

```
interface orsIImageSource : public orsIDataSource
{
public:
    virtual ors_uint32 getWidth()const = 0;
    virtual ors_uint32 getHeight()const = 0;

    virtual ors_uint getNumberOfInputBands()const = 0;
    virtual ors_uint getNumberOfOutputBands()const = 0;
```

```cpp
virtual orsDataTYPE getInputDataType(int iBand = 0)const = 0;
virtual orsDataTYPE getOutputDataType(int iBand = 0)const = 0;

//获取当前指定比例的影像窗口,若获取影像的比例与输入的比例不一致,则需要在读
//取后加入 Image Warper 节点,如 orsIImageZoomer
virtual orsIImageData * getImageData(orsRect_i &rect, double zoomRate, orsBand-
    Set &bandSet)=0;

//返回影像几何模型
virtual orsIImageGeometry * GetImageGeometry()= 0;
//返回波段元数据
virtual const orsIBandMetaData * getBandMetaData(int iBand)= 0;
//影像的范围,宽和高
virtual orsRect_i getBoundingRect(ors_uint band=0)const = 0;
...

ORS_INTERFACE_DEF(orsIDataSource, _T("image"))
};
```

影像源的数据类型保持与 GDAL 库给定的一致:

```cpp
/*! Pixel data types */
//与 GDAL 对应,可以直接转换
enum orsDataTYPE
{
    ORS_DT_UnKNOWN   = 0,
    ORS_DT_BYTE      = 1,/*! 8 bit unsigned Integer   */
    //ORS_DT_SINT8   = 2,/*! 8 bit signed Integer     */
    ORS_DT_UINT16    = 2,/*! 16 bit unsigned Integer  */
    ORS_DT_INT16     = 3,/*! 16 bit signed Integer    */
    ORS_DT_UINT32    = 4,/*! 32 bit unsigned Integer  */
    ORS_DT_INT32     = 5,/*! 32 bit signed Integer    */
    ORS_DT_FLOAT32   = 6,/*! 32 bit Floating point    */
    ORS_DT_FLOAT64   = 7,/*! 64 bit Floating point    */
    ORS_DT_CINT16    = 8,/*! Complex INT16            */
    ORS_DT_CINT32    = 9,/*! Complex INT32            */
    ORS_DT_CFLOAT32  = 10,/*! Complex FLOAT32         */
    ORS_DT_CFLOAT64  = 11,/*! Complex FLOAT64         */
```

```
    ORS_DT_COUNT    =  12,/* maximum type # + 1        */
};
```

若影像源具有几何成像模型(如仿射变换、共线方程、RPC 等),则 GetImageGeometry 返回几何成像模型接口指针,否则返回 NULL。

若影像源具有元数据(如 HJ-1A/1B、Landsat 7 等),则 getBandMetaData 返回元数据接口指针,否则返回 NULL。

4.1.2　影像处理链接口 orsIImageChain

影像链是 OpenRS 影像内存处理的主要方式,影像源的互相串接就是影像处理链。为了便于影像链的管理,专门定义了影像处理链对象。它的作用就是一个影像链的容器,能够方便地进行影像链的动态管理。实现影像源接口,orsIImageSource 的对象都可以作为节点加入影像链。图 4-1 为影像链的实现原理,若影像读取、变换、滤波、分割、分类都实现了 orsIImageSource 接口,则可以把这些操作加入影像链,从影像链对象获取数据就自动进行了一系列处理。

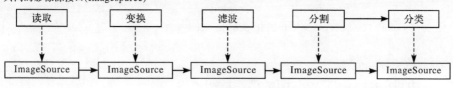

图 4-1　影像链的实现原理

影像处理链的基本操作包括影像源的添加、删除、插入、节点数获取、节点获取。

```
interface orsIImageChain :   public orsIImageSource
{
public:
    //加入影像链的尾部(right side)
    virtual bool add(orsIImageSource *object)= 0;
    virtual bool remove(orsIImageSource *object)= 0;
    virtual bool removeAll()= 0;

    //加入影像链节点的左侧
    virtual bool insertLeft (orsIImageSource * newObj, orsIImageSource * left-
        OfThisObj)= 0;
```

```
//加入影像链节点的右侧
virtual bool insertRight(orsIImageSource * newObj, orsIImageSource * right-
    OfThis-Obj)= 0;

//得到链条上所有的可连接对象
virtual orsIImageSource * getNode(int index)= 0;
virtual ors_uint getNumberOfNodes()= 0;

ORS_INTERFACE_DEF(orsIImageSource, _T("imageChain"))
};
```

4.1.3 影像服务接口 orsIImageService

OpenRS 是一个基于插件的系统，一般不知道对于一个具体的影像格式，是否存在一个能打开该影像的插件，因此 OpenRS 的影像读取是通过遍历影像读取对象来实现的。另外，为了方便常用影像对象的创建，这里设立了一个影像服务对象。接口如下：

```
interface orsIImageService : public orsIService
{
public:
    virtual int SizeOfType(orsDataTYPE type)= 0;

    //得到目前支持的所有的影像文件格式
    virtual orsFileFormatList getSupportedImageFormats()= 0;

    //打开一个影像文件，将遍历所有实现了 orsIImageReader 的对象来实现
    virtual orsIImageSourceReader * openImageFile(const orsChar * filename, bool
        bUpdate=false)= 0;

    //打开一个元数据文件，将遍历所有实现了 orsIImageMetaDataReader 的对象来实现
    virtual orsIImageMetaData * openImageMetaData(const orsChar * filename)= 0;

    virtual const orsChar * getPreDefinedResampleMode(orsResampleMODE resampleMode)= 0;

    virtual orsIImageData * CreateImageData(orsDataTYPE dataType)= 0;
    virtual orsIImageData * CreateImageData()= 0;
    virtual orsIImageSourceCache * CreateImageSourceCache()= 0;
```

```
    virtual orsIImageChain * CreateImageChain()= 0;
    virtual orsIAlgImageConvolution * CreateImageConvolution()= 0;

    //取影像源的统计信息,可以是变换之后的,modifier 为文件名修饰符,用于读取和保
    存统计信息
    virtual orsIAlgImageStatistics * CreateStatitics(orsIImageSource * pImg, bool
        bApproxOK)= 0;

    //影像读写缓存
    virtual orsIImageReadWriteCache * CreateImageReadWriteCache()= 0;

    //读取缩小影像时是否进行抽样, return previous state
    virtual bool enableApproximateRead(orsIImageSource * pImg, bool bEnable)= 0;

    ORS_INTERFACE_DEF(orsIService, _T("image"))
};
```

影像服务中最常用的就是 openImageFile、CreateImageData 和 openImageMetaData。openImageFile 用于打开一个已经存在的影像,CreateImageData 用于创建一个影像数据块,openImageMetaData 用于打开一个影像可能存在的元数据文件。

openImageFile 和 openImageMetaData 都采用遍历尝试影像读取对象和元数据读取对象的方式来实现影像和元数据的打开。

1. 打开影像

```
#include "orsBase/orsIRegisterService.h"
#include "orsBase/orsIPlatform.h"
#include "orsImage/ orsIImageSourceReader.h"

orsIImageSourceReader * orsXImageService::openImageFile(const char * fileName,
    bool bUpdate)
{
    orsIRegisterService * registerService = getPlatform()-> getRegisterService();

    //获取实现 orsIImageSourceReader 的所有对象
    orsArray<ref_ptr<orsIObjectDesc> > objDescs = registerService-> getObject-
        DescsByInterface("orsIImageSourceReader");
```

```cpp
//遍历所有对象
for(unsigned i=0;i< objDescs.size();i++)
{
    orsIObjectDesc *desc = objDescs[i].get();
    //创建该对象
    orsIImageSourceReader *reader = ORS_CREATE_OBJECT(orsIImageSourceReader, desc->getID());

    if(NULL != reader){
        //尝试打开
        if(reader->open(fileName, bUpdate))
            return reader;
        else
            reader->release();
    }
}
getPlatform()->logPrint(ORS_LOG_WARNING, "Can not open image %s", fileName);

return NULL;
}
```

2. 打开元数据

```cpp
#include "orsBase/orsIRegisterService.h"
#include "orsBase/orsIPlatform.h"
#include "orsImage/ orsIImageMetaDataReader.h"

orsIImageMetaData *orsXImageService::openImageMetaData(const orsChar *filename)
{
    orsIRegisterService *registerService = getPlatform()->getRegisterService();

    //获取实现 orsIImageMetaDataReader 的所有对象
    orsArray<ref_ptr<orsIObjectDesc> > objDescs =
        registerService->getObjectDescsByInterface("orsIImageMetaDataReader");
    …
    //遍历所有对象
```

```
        for(unsigned i=0; i < objDescs.size(); i++)
        {
            orsIObjectDesc *desc = objDescs[i].get();

            //创建该对象
            ref_ptr<orsIImageMetaDataReader> reader
                = ORS_CREATE_OBJECT(orsIImageMetaDataReader, desc->getID());

            if(NULL != reader.get()){
                //尝试打开
                orsIImageMetaData *pMetaData = reader->open(filename);
                if(NULL != pMetaData)
                    return pMetaData;
            }
        }
        return NULL;
    }
```

4.1.4 写影像接口

OpenRS 写影像接口 orsIImageWriter 用于创建给定格式的影像,并写入数据。目前最常用的实现是基于 gdal 的影像写入。除文件名外,影像的创建一般要指定大小、波段数、数据类型、成像模型和空间参考。为了简化影像的创建,我们设计了多个版本的影像函数,主要是利用参考影像的信息简化影像的创建。在给定的参考影像 infoFrom 的基础上,可以重新指定影像的波段数、数据类型、影像范围、影像大小等信息。在对输入影像进行处理,然后输出新的影像时,由于大部分的影像信息是不变的,因此利用合适的创建函数可以极大地简化创建的代码。

```
interface orsIImageWriter: public orsIImageSourceReader
{
public:
    ...

    //创建影像指定,影像信息从参考影像获得,波段数不为 0 则指定新的波段数
    virtual bool Create(const char *fileName, orsIImageSource *infoFrom, int numOfBands = 0) = 0;
    virtual bool Create(const char *fileName, orsIImageSource *infoFrom, orsDataTYPE dataType, int numOfBands) = 0;
```

```
//指定创建窗口范围
virtual bool Create(const char * fileName, orsIImageSource * infoFrom, const
    orsRect_i &rect, int numOfBands = 0)= 0;
virtual bool Create(const char * fileName, orsIImageSource * infoFrom, const
    orsRect_i &rect, orsDataTYPE dataType, int numOfBands)= 0;
virtual bool Create(const orsChar * fileName, orsIImageSource * infoFrom, int
    imgWidth, int imgHeight)= 0;
//创建影像指定,指定几何变换、空间参考等
virtual bool Create(const char * fileName, orsSIZE &imgSize, int bands, orsData-
    TYPE dataType, const geoImgCacheINFO * imgInfo, const char * pszProjection)
    = 0;

virtual bool WriteBandRect(int iBand, int x0, int y0, int nWid, int nHei, BYTE *
    buffer)= 0;
virtual void Close()= 0;

ORS_INTERFACE_DEF(orsIImageSourceReader, _T("writer"))
};
```

4.1.5 影像数据的读取与处理

OpenRS 的影像读取采取影像块的方式,为了统一管理生命周期,采取了对象的方式。从 orsIImageSource 返回的影像块接口指针,根据指针进行操作。采用影像块对象的好处是可以把影像块的完整信息传递给其他对象进行进一步处理。

1. 影像数据块接口

对于创建者,orsIImageData 需要指定影像块的数据类型,影像块的范围和波段集合调用 create 来初始化影像块。然后可以通过 setRange 来改变影像范围和波段集。

```
interface orsIImageData : public orsIObject
{
public:
    //创建
    virtual bool create(orsDataTYPE dataType, const orsRect_i &rect, const orsBandSet
        &bandSet)= 0;
```

```
    virtual void setZoomRate(double zoomRate)= 0;
    //设置范围和波段集合
    virtual void setRange(const orsRect_i &w, const orsBandSet &bandSet)= 0;

    //长度与宽度
    virtual ors_uint32 getWidth()const = 0;
    virtual ors_uint32 getHeight()const = 0;
    virtual const orsRect_i &getRect()const = 0;

    virtual orsDataTYPE getDataType(int iBand = 0)const = 0;

    virtual ors_uint getNumberOfBands()const = 0;
    virtual orsBandSet getBandSet()const = 0;
    virtual ors_byte * getBandBuf(int iBand = 0)const = 0;
    ...
    ORS_INTERFACE_DEF(orsIObject, _T("imageData"))
};
```

对使用者而言，主要从 orsIImageData 获取波段数据，通过 getBandBuf 返回字节数据。

2. 基于影像块的影像处理基本流程

为了说明从影像读取、影像处理到影像输出的完整过程，设计了以下代码。该代码先打开给定的影像，然后根据打开的影像，创建输出影像，处理时按照 256×256 大小对影像进行分块处理。为了简单起见，把影像数据转换成了浮点类型，并只对影像做了灰度减半处理。

```
#include "orsImage/orsIPlatform.h"
#include "orsImage/orsIImageService.h"
#include "orsImage/orsIImageWriter.h"

//打开影像
 ref_ptr<orsIImageSourceReader> imgReader = getImageService()-> openImageFile
    (imageFileName);

 if(NULL == imgReader.get())
    return false;
```

```
//创建写影像对象
ref_ptr<orsIImageWriter> imgWriter = ORS_CREATE_OBJECT(orsIImageWriter,
    _T("ors.dataSource.image.reader.writer.gdal"));
if(NULL == imgWriter.get())
    return false;


int nBands = imgReader->getNumberOfOutputBands();
//创建影像,因输出类型为浮点,这里要指定创建影像的数据类型
if(!imgWriter->Create (outputImageFileName, imgReader.get(), ORS_DT_FLOAT32,
    nBands))
    return false


float zoomRate = 1;
orsBandSet bandSet;
int nImgWid = imgReader->getWidth();
int nImgHei = imgReader->getHeight();

int iBand;
for(iBand=0; iBand < nBands; iBand++)
    bandSet.pushback(iBand);


orsRect_i rect;
const orsIImageData * imgData;
BYTE * bandBuf0 = new float[256 * 256], * bandBuf;

int row0, col0, row, col;
for(row0=0; row0 < nImgHei; row0+=256)
{
    rect.m_ymin = row0;
    rect.m_ymax = row0+256;

    if(rect.m_ymax > nImgHei)
        rect.m_ymax = nImgHei;

    for(col0=0; col0 < nImgWid; col0+=256)
    {
        rect.m_xmin = col;
        rect.m_xmax = col+256;
```

```
        if(rect.m_xmax > nImgWid)
            rect.m_xmax = nImgWid;

        //读取影像块
        imgData = imgReader-> getImageData(rect, zoomRate, bandSet);

        for(iBand=0; iBand < nBands; iBand++)
        {
            //波段数据转换成浮点类型
            imgData-> buf2float(bandBuf0);
            bandBuf = bandBuf0;
            for(row=0; row < rect.Height(); row++)
            {
                for(col=0; col < rect.Width(); col++)
                {
                    *bandBuf++ /= 2.0;
                }
            }

            //写出修改后的波段数据
            imgWriter-> WriteBandRect(iBand, col0, row0, rect.Width(), rect.
                Height(), (BYTE *)bandBuf0);
        }
    }
}

delete bandBuf;
```

4.2 影像几何处理模块 orsImageGeometry

  遥感数据处理与一般的图像处理不同,除了一般的图像处理,遥感数据处理还要关注数据的几何信息和辐射信息。因此,几何和辐射相关处理是 OpenRS 的技术核心之一。OpenRS 的几何处理从遥感数据的应用出发,定义合理的几何模型接口,通过通用的模型屏蔽具体的成像原理和模型,以实现数据的扩展。OpenRS 的几何处理不包括遥感数据预处理的内容。

4.2.1 遥感应用中影像几何处理的需求

遥感应用中的几何处理一般都和影像的演示、几何纠正、立体建模等处理相关，例如：

(1) 不同坐标系、不同比例尺的影像如何叠加显示？
(2) 正射影像、原始影像能否叠加显示？
(3) 任意两个影像如何进行几何匹配或配准？
(4) 不同传感器的影像如何进行统一的立体显示？
(5) 不同传感器的影像如何进行统一的密集匹配生成 DSM？
……

下面分析不同传感器影像的成像方式。

如图 4-2(a)、(b)所示，目前的航空影像主要分两种，即面阵中心投影成像和线阵推扫式多中心投影。例如，DMC、UCD/UCX 等航空数码相机为面阵中心投影；ADS40/80 为线阵推扫式多中心投影。

如图 4-2(c)所示，卫星影像的成像多为线阵推扫式中心投影（江恒彪等，2009)，如 ALOS、SPOT、IKONOS/GeoEye、QuickBird。

如图 4-2(d)所示，SAR 影像则为距离成像（王淑艳等，2011）。

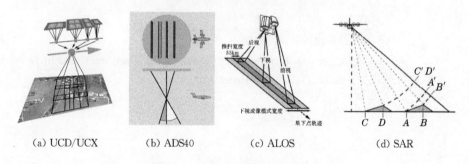

(a) UCD/UCX   (b) ADS40   (c) ALOS   (d) SAR

图 4-2 不同传感器的几何模型

从以上分析可以看出，遥感应用中遥感影像几何处理的核心问题是"不同传感器、坐标系、比例尺的影像统一处理问题。"

4.2.2 成像几何模型的统一表示

根据前面的分析，OpenRS 采用通用成像模型统一地表示光学、SAR、红外等各种形式的影像几何模型。基于通用成像模型进行影像方位元素的精化、影像的几何纠正将能够极大地简化软件的设计、提高影像的纠正速度，并有助于提高软件的可靠性。

严格模型的统一表示由接口来定义,不同影像的成像模型由不同的对象来实现。基于该接口,几何纠正、单片测图等应用程序可以不需要关注具体的成像模型。图 4-3 给出了 OpenRS 的几何模型对象体系。

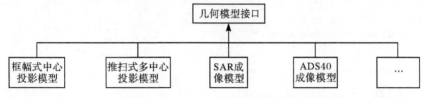

图 4-3 OpenRS 几何模型的对象体系

4.2.3 遥感影像的基本几何处理

遥感影像的几何处理主要关心影像坐标及地面坐标的转换,一般包括投影与交会两类(图 4-4)。

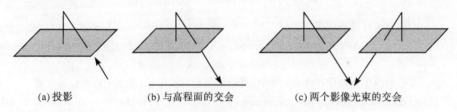

图 4-4 典型的影像几何操作

1) 投影计算

投影计算是遥感影像成像模型的核心接口,提供由地面物体坐标计算该点在影像上成像位置的功能。

$$(x,y) \leftarrow (X,Y,Z)$$

2) 光束与高程面的交会

光束与高程面的交会可以确定光束在不同高程时的 X、Y 坐标,可以由计算给定影像坐标的平面位置。

$$(X,Y) \leftarrow (x,y,Z)$$

3) 光束与光束的交会

光束与光束的交会是立体摄影测量中必备的核心功能之一。能够用于立体匹配。由于涉及两个影像,因此不属于成像模型本身的功能。

$$(X,Y,Z) \leftarrow (x_1,y_1;x_2,y_2)$$

4.2.4 影像几何模型接口

除了投影和交会两个接口,为了实现立体交会,对于非线性模型,影像几何模型还需提供偏微分、偏导数接口以辅助实现多个光束的立体交会计算。

OpenRS 的影像几何模型接口 orsIImageGeometry 如下：

```
struct orsIImageGeometry : public orsIObject
{
    virtual ~orsIImageGeometry(){};

    virtual intGetOrder()const = 0;
    //影像光束
    //1. 像点处的成像光束
    virtual bool RayVector(double xi, double yi, double Z, double *a, double *b,
        double *c)= 0;
    //2. 过点物方点的成像光束
    virtual bool RayVector_3D(double X, double Y, double Z, double *a, double *b,
        double *c)= 0;

    //物方坐标投影到像方坐标
    virtual void Project(double X, double Y, double Z, double *x, double *y)= 0;

    //获取给定点处的线性光束矩阵,并设定当前像点坐标,像点坐标单位为像素
    virtual void GetLinearMatrix(double x, double y, orsMatrixD &mA, orsVectorD &vL)
        = 0;
    //获取给定点处成像模型对 XYZ 的偏导数,当前像点在 GetLinearMatrix 中设定
    virtual void PartialDerivative_XYZ(double xg, double yg, double zg, orsPD_PIXEL_
        XYZ *pd)= 0;
    //获取给定点处前方交会矩阵,当前像点在 GetLinearMatrix 中设定
    virtual void SpaceIntersectMatrix(double xg, double yg, double zg, orsMatrixD
        &mA, orsVectorD &vL)= 0;
    //影像光束与物方三维表面或直线的交会
    //1. 影像光束与高程面的交点,正射影像的交会值等于变换后的物值
    virtual void IntersectWithZ(double x, double y, double Z, double *X, double *Y)
        = 0;
    //2. 影像光束与垂线的交点
    virtual bool IntersectWithXY(double xi, double yi, double X, double Y, double *
        Z)= 0;
    //3. 影像光束与任意直线的交点
    virtual double IntersectWithLine(double xi, double yi, const orsLINE3D &line)
        = 0;
```

```cpp
//设置任意交会平面
virtual void SetIntersectPlane(const orsPLANE *plane)= 0;
//4. 光束与平面交点
virtual bool IntersectWithPlane(double xi, double yi, orsPOINT3D *point)= 0;
//5. 交点对平面角度的导数
virtual bool PartialDerivative_Plane(double xi, double yi, orsPD_XYZ_PLANE *
    pd, orsPOINT3D *point)= 0;

//判断影像上的线段在物方是否是垂线
virtual bool IsVertical(double xi[2], double yi[2], double distTh)= 0;

//取给定高程面上的地面采样间隔
virtual float GetPixelGSD(double Z = 0)= 0;
virtual float GetResolution_X(double Z=0)= 0;
virtual float GetResolution_Y(double Z=0)= 0;

//取给定高程的高程分辨率
virtual float GetResolution_Z(double Z = 0)= 0;

//取高程范围
virtual double GetMeanZ()const = 0;
virtual double GetMinZ()const = 0;
virtual double GetMaxZ()const = 0;

//飞行高度
virtual double GetFlyHeight()const = 0;

//设置物方坐标系统,若设置成功,则程序内部自动对物方坐标进行坐标系转换
virtual bool SetSRS(const char *hcsWkt, const char *vcsWkt)= 0;

//获取空间参考
virtual orsISpatialReference *GetSpatialReference()= 0;

public:
    ORS_INTERFACE_DEF(orsIObject, _T("imageGeometry"));
};
```

4.2.5 多个影像光束的交会

多个影像光束的交会接口如下。该接口定义了影像交会的功能，即添加影像光束并进行影像交会，另外还提供光束夹角计算功能（图4-5）。

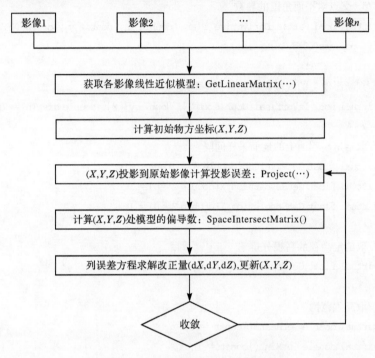

图4-5 多个影像光束的空间交会过程

```
struct orsISpaceIntersection : public orsIAlgorithm
{
public:
    virtual ~orsISpaceIntersection(){};

    //可以多至8个影像
    virtual bool AddImageRay(orsIImageGeometry *pImgRay)= 0;
    virtual bool RemoveImageRay(orsIImageGeometry *pImgRay)= 0;

    //计算交会点
    virtual void Intersect(double *x, double *y, double *X, double *Y, double *
        Z)= 0;
    //估算交会角
```

```
    virtual float GetIntersectAngle(double *x, double *y)= 0;


public:
    ORS_INTERFACE_DEF(orsIAlgorithm, "spaceIntersect")
};
```

4.2.6 空间参考与坐标变换 orsSRS

空间参考描述了一个地物在地球上的真实位置。为了正确地对位置进行描述，需要引入一个可供测量和计算的框架，使得大地测量的结果能够在这个框架上进行描述。地球是一个不规则形状的椭球体，那么使用什么样的方法来模拟地球的形状，又该如何将球面上的坐标投影在平面的地图上？这就需要先了解大地水准面、参考椭球体、基准面的概念，以及它们之间的关系。本书不准备讨论这些具体的内容，只是给出 OpenRS 中进行相关计算的软件接口。

OpenRS 的空间参考模块是基于 gdal 库的 OGR 模块和 proj 库实现的。空间参考模块的接口包括地球椭球接口、空间参考接口和坐标变换接口。

1. 椭球接口

椭球接口用于设置椭球的长轴和短轴，并提供调用该椭球上地心直角坐标和大地坐标（即经纬度及椭球高）之间的变换函数。

```
interface orsIEllipsoid : public orsIObject
{
public:
    virtual bool SetParameter(double a, double b)= 0;

    virtual double GetSemiMajor()const = 0;
    virtual double GetSemiMinor()const = 0;

    virtual bool Geodetic_To_Geocentric (double latitude, double longitude, double
        height, double *X, double *Y, double *Z)= 0;
    virtual bool Geocentric_To_Geodetic (double X, double Y, double Z, double *lati-
        tude, double *longitude, double *height)= 0;

    ORS_INTERFACE_DEF(orsIObject, _T("orsIEllipsoid"));
};
```

2. 空间参考接口

空间参考接口用于设定一个空间参考的参数及调用相应的坐标变换。导入类函数（import 开头）用于从特定格式的参数、字符串或编码对空间参考进行初始化。导出类函数（export 开头）用于导出特定格式的空间参考参数、字符串或编码。坐标变换类函数用于该空间参考下大地坐标与地心坐标之间的转换；如果是投影坐标系，还可以对投影坐标和大地坐标进行转换。

```
interface  orsISpatialReference : public orsIObject
{
public:
    orsISpatialReference(){};
    virtual ~orsISpatialReference(){};

    //投影坐标与地心坐标的转换
    virtual bool Projected_To_Geodetic(double x, double y, double *latitude, double
        *longtitude)= 0;
    virtual bool Geodetic_To_Projected(double latitude, double longtitude, double
        *x, double *y)= 0;

    //大地坐标与地心坐标的转换
    virtal bool Geodetic_To_Geocentric (double latitude, double longitude, double
        height, double *X, double *Y, double *Z)= 0;
    virtual bool Geocentric_To_Geodetic (double X, double Y, double Z, double *lati-
        tude, double *longitude, double *height)= 0;

    //地心坐标与 WGS84 地心坐标的转换
    virtual bool Geocentric_To_WGS84(double *x, double *y, double *z)= 0;
    virtual bool Geocentric_From_WGS84(double *x, double *y, double *z)= 0;

public:
    virtual bool SetWithSpatialReference(orsISpatialReference *pSpatialReference)
        = 0;

    //设置高程系统
    virtual bool SetVcs(vcsTYPE vcsType, const char *vcsWkt = NULL)= 0;

public:
```

//坐标系比较
virtual bool IsSameSRS(orsISpatialReference * toCS)const = 0;
virtual bool IsSameOgrCS(orsISpatialReference * toCS)const = 0;
virtual bool IsSameOgrGeogCS(orsISpatialReference * toCS)const = 0;

virtual bool IsProjected()const = 0;
virtual bool IsGeographic()const = 0;
virtual double GetSemiMajor()const = 0;
virtual double GetSemiMinor()const = 0;

virtual vcsTYPE VcsType()const = 0;
virtual orsIGeoid * GetGeoid()= 0;
//参数的导入
virtual bool importFromOGR(const OGRSpatialReference * pOgrSR)= 0;

virtual bool importFromWkt(const char * hcsWkt)= 0;
virtual bool importFromVcsWkt(const char * vcsWkt)= 0;

virtual bool importFromProj4(const char * pszProj4)=0;
virtual bool importFromEPSG(int epsgCode)=0;
virtual bool importFromESRI(const char * papszPrj)=0;
virtual bool importFromPCI(const char * pszProj, const char * pszUnits = NULL,
    double * padfPrjParams = NULL)=0;
vortual bool importFromUSGS(long iProjSys, long iZone, double * padfPrjParams,
    long iDatum)=0;
virtual bool importFromPanorama(long iProjSys, long iDatum, long iEllips, long
    iZone, double dfStdP1, double dfStdP2, double dfCenterLat, double dfCenter-
    Long)=0;
virtual bool importFromXML(const char * pszXML)=0;
virtual bool importFromDict(const char * pszDict, const char * pszCode)=0;

virtual bool SetProjection(const char *)= 0;
virtual bool SetGeogCS(const char * pszGeogName, const char * pszDatumName,
    const char * pszEllipsoidName, double dfSemiMajor, double dfInvFlattening,
    const char * pszPMName = NULL, double dfPMOffset = 0.0, const char * psz-
        Units = NULL, double dfConvertToRadians = 0.0)= 0;
//参数的导出
virtual const OGRSpatialReference * GetOGRSpatialReference()= 0;

```cpp
    virtual bool exportToWkt(ref_ptr<orsIOGRString> &hcsWkt)const = 0;
    virtual bool exportToVcsWkt(ref_ptr<orsIOGRString> &vcsWkt)const = 0;
    virtual bool exportToPrettyWkt(ref_ptr<orsIOGRString> &ppszResult, bool bSim-
        plify = false)const = 0;
    virtual bool exportToProj4(ref_ptr<orsIOGRString> &ppszProj4)const = 0;
    virtual bool exportToPCI(ref_ptr<orsIOGRString> &ppszProj, ref_ptr<orsI-
        OGRString> &ppszUnits, ref_ptr<orsIOGRValueArray> &ppadfPrjParams)const
        = 0;
    virtual bool exportToUSGS(long *piProjSys, long *piZone, ref_ptr<orsIOGRVal-
        ueArray> &ppadfPrjParams, long *piDatum)const = 0;
    virtual bool exportToXML(ref_ptr<orsIOGRString> &ppszRawXML, char *pszDialect
        = NULL)const = 0;
    virtual bool exportToPanorama(long *piProjSys, long *piDatum, long *piEl-
        lips, long *piZone, double *pdfStdP1, double *pdfStdP2, double *pdfCen-
        terLat, double *pdfCenterLong)const = 0;

    virtual bool SetTOWGS84(double, double, double, double = 0.0, double = 0.0,
        double = 0.0, double = 0.0)=0;
    virtual bool GetTOWGS84(double *padfCoef, int nCoeff = 7)const = 0;

    ORS_INTERFACE_DEF(orsIObject, _T("spatialReference"))
};
```

3. 坐标变换接口

坐标变换接口用于两个空间参考坐标系之间的坐标变换，包括正变换 Transform 和逆变换 Inverse。通过初始化给定范围的近似变换，可以简化变换过程，提高计算的速度，特别适合于大批量的坐标变换。InitFastTransform 函数给定参加变换的两个空间参考，InitFastTransform 用于初始化快速变换。

```cpp
interface orsICoordinateTransform: public orsIObject
{
public:
    orsICoordinateTransform(){};
    virtual ~orsICoordinateTransform(){};

    virtual void Initialize(orsISpatialReference *from, orsISpatialReference *to)
```

```cpp
        = 0;
    virtual bool InitFastTransform(double minX, double minY, double maxX, double
        maxY, double maxErr)= 0;
    virtual void EnableFast(bool bEnable = true)= 0;

    //正变换
    virtual long Transform(const orsPOINT3D *ptsFrom, int n, orsPOINT3D *ptsTo)
        = 0;
    //反变换
    virtual long Inverse(const orsPOINT3D *ptsFrom, int n, orsPOINT3D *ptsTo)= 0;
    //便利函数
    long Transform(const orsPOINT3D &ptFrom, orsPOINT3D *ptTo)
    {
        return Transform(&ptFrom, 1, ptTo);
    };
    long Transform(orsPOINT3D *pts, int n = 1)
    {
        return Transform(pts, n, pts);
    };
    long Inverse(const orsPOINT3D &ptFrom, orsPOINT3D *ptTo)
    {
        return Inverse(&ptFrom, 1, ptTo);
    };
    long Inverse(orsPOINT3D *pts, int n = 1)
    {
        return Inverse(pts, n, pts);
    };

    ORS_INTERFACE_DEF(orsIObject, _T("CoordinateTransform"))
};
```

4.2.7 动态影像几何变换——imageSourceWarper

imageSourceWarper 是影像变形（或变换）处理节点，用于建立两个影像之间的透明变换。在建立影像变形模型之后，对变换后影像块的处理就可以忽略变换的过程和细节，而专注于后续处理代码的编写。已经实现的 imageSourceWarper 包括影像放大 orsIImageSourceZoomer、影像旋转 orsIImageSourceRotator、相对纠正 orsIImageSourceWarper_refImg、成像模拟 orsIImageSourceWarper_imaging、正射纠正 orsIImageSourceWarper_ortho、多项式变换 orsIImageSourceWarper_polynomial 等。这部分功能以插件 orsImageSourceWapper.dll 的形式提供。

动态影像几何变换接口 orsIImageSourceWarper 及其派生类继承自影像源接口 orsIImageSource 和可链接接口 orsIConnectableObject。在使用时需要使用 connect 函数连接要被变形的影像源。

1. 影像重采样 orsIAlgImageResampler

影像重采样是影像变形等处理的重要环节。为了提高影像变形的效率，除标准的单点重采样外，影像重采样接口 orsIAlgImageResampler 考虑了影像块的变换采样。在 orsIAlgImageResampler 中，interpolate 是点采样接口、zoomRect 是影像块缩放接口，warper_affine2D 是二维仿射变换接口，warper_bilinear 是双线性变换接口，warper_dlt2D 是二维直接线性变换（又称为同形变换）接口。这样可以通过调用一个函数实现这四种常用的块变换，每一种重采样算法都需要实现这样的接口。

```
interface orsIAlgImageResampler:   public orsIAlgorithm
{
public:
    //取采样核的大小
    virtual void setDegree(ors_int32 splineDegree)= 0;
    virtual ors_int32 getKernelSize()= 0;

    //设置要重采样的影像
    virtual bool setImage(const orsIImageData * inputImg)= 0;

    //按起点和比例,内插一个影像
    //用于影像缩放,此时先调用 SetImage 设置适当的采样影像
    //不同尺度的影像错位调整自动进行
    virtual void zoomRect(orsIImageData * outImg)= 0;

    //不同尺度的影像错位调整必须包含在变换参数中
    //内插一个点的值,必须先调用 SetImage
    virtual void interpolate(double x, double y, void * buf)= 0;

    virtual void warp_dlt2D(orsIImageData * outImg, double * l)= 0;
    virtual void warp_affine2D(orsIImageData * outImg, double * a, double * b)= 0;
```

```
    virtual void warp_bilinear(orsIImageData *outImg, double *a, double *b)= 0;

    ORS_INTERFACE_DEF(orsIAlgorithm, "imageResampler")
};
```

其中,仿射变换、双线性变换、直接线性变换如下。

(1) 仿射变换:
$$\begin{cases} x' = a_0 + a_1 x + a_2 y \\ y' = b_0 + b_1 x + b_2 y \end{cases}$$

(2) 双线性变换:
$$\begin{cases} x' = a_0 + a_1 x + a_2 y + a_3 xy \\ y' = b_0 + b_1 x + b_2 y + b_3 xy \end{cases}$$

(3) 直接线性变换:
$$\begin{cases} x' = \dfrac{l_0 x + l_1 y + l_2}{l_6 x + l_7 y + l_8} \\ y' = \dfrac{l_3 x + l_4 y + l_5}{l_6 x + l_7 y + l_8} \end{cases}$$

影像重采样的算法很多,OpenRS 内置实现的包括最近邻、双线性、双三次和六点升余弦(raised cosine)四种。

```
#define ORS_ALG_IMAGEREAMPLER_NEAREST    "ors.algorithm.imageResampler.nearest"
#define ORS_ALG_IMAGEREAMPLER_BILINEAR   "ors.algorithm.imageResampler.bilinear"
#define ORS_ALG_IMAGEREAMPLER_BICUBIC    "ors.algorithm.imageResampler.bicubic"
#define ORS_ALG_IMAGEREAMPLER_RC6P       "ors.algorithm.imageResampler.rc6p"
```

2. 影像缩放器 orsIImageSourceZoomer

影像缩放器用于实现非整数倍的影像缩放。缩放参数 zoomRate 由 orsIImageSource 的 getImageData 给出。

```
interface orsIImageSourceZoomer : public orsIImageSource
{
public:
    virtual void setResampleMode(orsResampleMODE resampleMode, int sampleDegree =
        3)= 0;
    virtual void setResampleMode(const orsString &resampleMode, int sampleDegree =
        3)= 0;
```

```
        ORS_INTERFACE_DEF(orsIImageSource, "zoomer")
};
```

3. 影像旋转器 orsIImageSourceRotator

影像旋转器用于对影像进行动态旋转。

```
enum orsRoateMODE{
    ORS_rmNONE = 0,//不做旋转
        ORS_rmANTICLOCKWISE90, //逆顺时针旋转 90 度
        ORS_rmCLOCKWISE90, //顺时针旋转 90 度
        ORS_rm180, //180 度
        ORS_rmANY, //逆时针任意
};

class orsIImageSourceRotator: public orsIImageSource
{
public:
    virtual bool setRotateMode(orsRoateMODE rotateMode, ors_float64 rotateAngle =
        0)= 0;
    ORS_INTERFACE_DEF(orsIImageSource, "rotator")
};
```

4. 影像相对纠正 orsIImageSourceWarper_refImg

影像缩放和旋转比较好理解，影像相对纠正是为不同空间参考、不同几何模型影像之间进行整体对准定义的一个变换。影像相对纠正的基本参数为两个带有几何模型的影像源和一个高程面。如图 4-6 所示，根据影像几何模型接口可以计算出影像上一个窗口四个角与高程面的交点，然后投影到另一个影像上来确定两个影像窗口之间的关系，从而进行影像块的重采样。

```
class orsIImageSourceWarper_refImg: public orsIImageSourceWarper
{
public:
    //2D warping of {ORS_wmAFFINE, ORS_wmPROJECTIVE, ORS_wmBILINEAR}
    virtual bool setWarpMode(orsWarpMODE warpMode)= 0;

public:
```

```
//设置参考影像
virtual void setRefImage(orsIImageSource * pRefImage, float zRef)= 0;
//获取参考影像
virtual void getRefImage(orsIImageSource * * ppRefImage, float * zRef)= 0;
//设置纠正平面的高程
virtual void setRefPlane(float zRef)= 0;
//仅用于内插高程,不影响变换
virtual void attachDemForZ(orsIImageSource * pDem)= 0;

public:
    ORS_INTERFACE_DEF(orsIImageSourceWarper, "refImage")
};
```

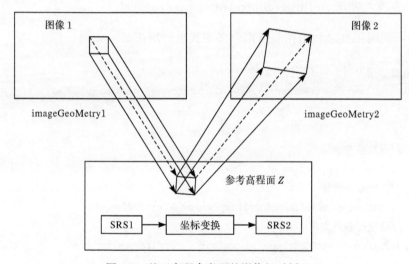

图 4-6 基于高程参考面的影像相对纠正

5. 影像仿真 orsIImageSourceWarper_imaging

在 orsIImageSourceWarper_refImg 中,高程面是一个固定的值,这样的影像变形只起到一个整体的对准作用。真正的对准需要考虑地形的起伏,以实现逐像素的一一对应。orsIImageSourceWarper_imaging 称为影像仿真接口,因为给定一个影像和对应的 DEM,可以生成一个与参考具有相同分辨率和成像角度的影像,从而为精细配准等应用提供了一个很好的影像变形改正工具。

```
//通过给定的原始影像和数字表面模型把正射影像模拟成原始影像
class orsIImageSourceWarper_imaging : public orsIImageSourceWarper
```

```
{
public:
    //设置原始影像和DSM
    virtual void setRefImage(orsIImageSource * pRefImage, orsIImageSource * dsm)
        = 0;
    //获取参考影像
    virtual orsIImageSource * getRefImage()= 0;

public:
    ORS_INTERFACE_DEF(orsIImageSourceWarper, "imaging")
};
```

6. 多项式纠正 orsIImageSourceWarper_polynomial

多项式纠正直接利用控制点拟合多项式进行纠正。

```
//通过给定控制点纠正到参考影像
class orsIImageSourceWarper_polynomial: public orsIImageSourceWarper
{
public:
    //设置参考影像
    virtual void setRefImage(orsIImageSource * pRefImage)= 0;
    //获取参考影像
    virtual void getRefImage(orsIImageSource * * ppRefImage)= 0;
    //设置控制点文件
    virtual void setGCPFile(const orsChar * pGcpFileName)= 0;

public:
    ORS_INTERFACE_DEF(orsIImageSourceWarper, _T("polynomial"))
};
```

4.3 影像元数据处理模块

虽然不同卫星传感器的影像往往具有不同的元数据文件格式,但是这些元数据的内容可以进行归类,以便定义出合适的接口,从而简化不同卫星影像数据的处理。将不同传感器的元数据读取包装成为插件的形式,以方便在统一的系统框架下进行扩充。

4.3.1 不同传感器影像的元数据

影像元数据自动读取由接口统一表示。卫星影像的元数据主要包括卫星平台、成像传感器、波段及原始影像信息四部分(唐游,2012)。每一部分都由相应的接口来表示(图 4-7)。

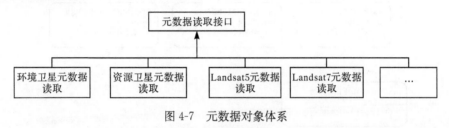

图 4-7 元数据对象体系

卫星平台元数据包括平台名称和平台类型。成像传感器元数据信息包括传感器名称、传感器类型、传感器成像方式、分辨率、幅宽、波段数及指定波段的波长信息。影像波段元数据包括波段号、成像传感器、原始影像信息、辐射定标参数、太阳天顶角和方位角、观测天顶角和方位角、分辨率、几何精度、地图投影、影像四个角的地图坐标和地理坐标及中心点的经纬度。原始影像的元数据包括传感器元数据、波段元数据、成像时间及产品的几何级别和辐射级别。

通过波段元数据接口可以获取传感器和原始影像的元数据接口,通过原始影像的元数据接口也可以得到平台和波段的元数据接口。

如果要支持新的卫星元数据读取,只需要实现元数据读取接口、卫星平台元数据接口、成像传感器元数据接口、波段元数据接口及原始影像元数据接口。

无论环境卫星、资源卫星还是其他卫星的元数据,对用户和算法开发者来说,元数据读取方式都一样。打开影像时,会自动读取该影像的元数据,并判断是否是资源卫星、环境卫星、陆地卫星 5 号或者陆地卫星 7 号的元数据,若是其中一种卫星的元数据,则元数据读取成功,否则失败,返回 NULL。

4.3.2 元数据接口设计

根据遥感影像数据获取的条件,把遥感数据的元数据接口分为对地观测平台、影像元数据和波段元数据三类(图 4-8)。对地观测平台描述了传感器的承载平台和飞行规律、波段元数据描述了波段数据直接相关的元数据,影像元数据描述了影像的波段构成、处理级别等信息。其中,波段元数据是直接面向图像处理使用的接口。

影像传感器用于获取影像数据的载荷,决定了影像数据的特性。根据传感器获取的信息类型可以分为反射可见光、发射不可见光的被动式传感器和主动式的

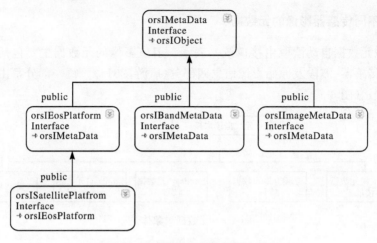

图 4-8　遥感数据的元数据接口分类

SAR 传感器(图 4-9)。通常,一个观测平台上可能不止一个影像传感器,一个影像可能包含多个不同传感器影像波段。

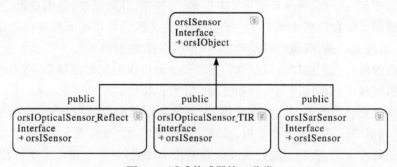

图 4-9　遥感传感器接口分类

4.3.3　传感器

　　传感器接口主要返回传感器的名称、传感器的类型及类型名称、传感器的数据获取模式、地面分辨率、成像宽度等参数。传感器类型和成像模式主要参考 Geographic information/Geomatics(ISO/TC 211)标准。

```
//传感器模式
enum orsSensorMODE {
    ORS_SM_UNKNOWN,
    ORS_SM_FRAMING,
    ORS_SM_PUSHBROOM,
```

```
    ORS_SM_SPOT,
    ORS_SM_SWATH,
    ORS_SM_WHISKBROOM,
    ORS_SM_BUMPER
};

//影像类型
enum orsSensorTYPE{
    ORS_ST_UNKNOWN,
    ORS_ST_PAN,     //反射可见光:全色
    ORS_ST_MS,      //反射可见光:multispectral
    ORS_ST_HS,      //反射可见光:hyperspectral
    ORS_ST_NIR,     //反射不可见光:near infrared
    ORS_ST_TIR,     //发射不可见:thermal infraRed
    ORS_ST_RADAR    //主动反射:RADAR
};

interface orsISensor : public orsIObject{
public:
    virtual ors_string getSensorID()= 0;        //传感器名称,如 PRISM
    virtual ors_int32 getSensorType()= 0;       //传感器类型
    virtual ors_string getSensorTypeName()= 0;  //传感器类型名称
    virtual orsSensorMODE getSensorMode()= 0;   //成像模式
    virtual ors_float32 getGSD()= 0;            //地面分辨率(m)
    virtual ors_float32 getSwathWidth()= 0;     //扫描宽度(m)

public:
    ORS_INTERFACE_DEF(orsIObject, _T("sensor"))
};
```

1. 光学传感器

光学传感器分为可见光传感器和红外传感器两类。orsBand 定义波段的波长范围。

```
struct orsBAND{
    ors_float32 startWaveLength;//微米
    ors_float32 endWaveLength;//微米
};
```

```cpp
interface orsIOpticalSensor_Reflect : public orsISensor{
public:
    virtual ors_int32 getNumOfBands()= 0;
    virtual orsBAND &getBand(int index)= 0;
    //波段平均太阳辐照度
    virtual ors_float32 getBandESUN(int index)= 0;

public:
    ORS_INTERFACE_DEF(orsISensor, _T("optical"))
    };

    interface orsIOpticalSensor_TIR : public orsISensor{
public:
    virtual ors_int32 getNumOfBands()= 0;
    virtual orsBAND &getBand(int index)= 0;

public:
    ORS_INTERFACE_DEF(orsISensor, _T("optical"))
};
```

2. SAR 传感器

SAR 传感器的元数据包括波长、频率和极化模式。

```cpp
enum orsPolarizeMODE{
    ORS_PLM_HH,
    ORS_PLM_VV,
    ORS_PLM_HV,
    ORS_PLM_VH
};

interface orsISarSensor : public orsISensor{
public:
    virtual ors_float32 getWaveLength()= 0;
    virtual ors_float32 getFrequency()= 0;

    virtual ors_int32 getNumOfPolarizeModes()= 0;
    virtual ors_int32 getPolarizeMode(ors_int32 index)= 0;
```

```cpp
public:
    ORS_INTERFACE_DEF(orsISensor, _T("sar"))
};
```

4.3.4 观测平台

观测平台分为航空器、卫星和气球三类。

```cpp
enum orsPlatformTYPE{
    ORS_PT_AIRCRAFT,
    ORS_PT_SATELLITE,
    ORS_PT_BALLON
};

interface orsIEosPlatform : public orsIMetaData{
public:
    virtual ors_string getPlatformID()= 0;//平台名称,如 ALOS
    virtual ors_string getPlatformTypeName()= 0;

    virtual orsPlatformTYPE getPlatformType()= 0;

public:
    ORS_INTERFACE_DEF(orsIMetaData, _T("eosPlatform"))
};
```

卫星观测平台的元数据包括轨道参数(轨道高度、轨道倾角、轨道周期)、重返周期、传感器。

```cpp
structorsORBIT{
    ors_float32 altitude;//轨道高度(km)
    ors_float32 inclination;//轨道倾角(°)
    ors_float32 period;//轨道周期(min)
};

interface orsISatellitePlatfrom : public orsIEosPlatform{
public:
    virtual ors_float32 getOrbitAltitude()= 0;
    virtual ors_float32 getOrbitInclination()= 0;
```

```
    virtual ors_float32 getOrbitPeriod()= 0;

    virtual ors_float32 getRepeatCycle()= 0;//重返周期(d)

    virtual ors_int32 getNumOfSensors()= 0;
    virtual orsISensor * getSensor(ors_int32 index)= 0;

public:
    ORS_INTERFACE_DEF(orsIEosPlatform, _T("satellite"))
};
```

4.3.5 影像元数据

1. 影像元数据

影像元数据指与影像获取相关的所有参数,包括传感器平台、波段数、波段元数据、获取时间、产品级别等相关信息。

```
interface orsIImageMetaData : public orsIMetaData{
public:
    ///////////////////////成像平台与传感器/////////////////////////
    virtual orsIEosPlatform * getSensorPlatform()= 0;

    virtual ors_int32 getNumOfBands()= 0;//波段数,多波段文件
    //根据波段文件名,判断波段号,从1开始,失败返回-1
    virtual int getBandNum(const orsChar * fileName)= 0;
    virtual orsIBandMetaData * GetBandMetaData(int iBand)= 0;
    virtual const orsIProperty * GetMetaReaderProperty ()const =0 ;

    ///////////////////////原始成像信息/////////////////////////
    //获取时间
    virtual bool getImagingDate(orsDataDATE &date)= 0;

    //产品级别信息
    virtual orsImageGeometryLEVEL getGeometryLevel()= 0;
    virtual orsImageRadianceLEVEL getRadianceLevel()= 0;

    virtual void setImageSource(orsIImageSource * pImg)= 0;
```

```
public:
    ORS_INTERFACE_DEF(orsIMetaData, _T("image"))
};
```

2. 波段元数据

波段元数据包括波段 ID、分辨率、地面覆盖范围、原始成像信息、辐射定标系数（增益和偏置）、像元的太阳天顶角和方位角、像元的传感器天顶角和方位角等及该波段影像获取的相关所有信息。

```
interface orsIBandMetaData : public orsIMetaData
{
public:
    virtual ors_string getBandID()= 0 ;
    virtual orsISensor * getSensor()= 0 ;

    //原始成像信息
    virtual orsIImageMetaData * getImageMetaData()= 0;

    //辐射定标信息
    virtual double getGainOffset()=0;
    virtual double getGain()= 0;

    virtual bool hasPerPixelSunAngle()= 0;
    virtual bool hasPerPixelSensorAngle()= 0;
    //像元对应的太阳天顶角、方位角
    virtual bool getSunAngle(int row, int col, ors_float32 * sunZenith, ors_float32
        * sunAzimuth)= 0 ;
    //像元对应的的观测天顶角、方位角
    virtual bool getSensorAngle(int row, int col, ors_float32 * sensorZenith, ors_
        float32 * sensorAzimuth)= 0;
    //与波段对应的观测天顶角、方位角影像.
    virtual orsIImageSource * getSensorAngleImg()= 0 ;

    //分辨率
    virtual ors_float32 getGSD_Row()= 0;
    virtual ors_float32 getGSD_Col()= 0;
```

```
//几何精度
virtual ors_float32 getCE90_XY()= 0;
virtual ors_float32 getCE90_Z()= 0;//立体产品

//投影坐标
virtual bool getMapProjection(orsString * wktStr)=0;

virtual bool getTopLeftMapXY(orsPOINT2D * pt)= 0;
virtual bool getTopRightMapXY(orsPOINT2D * pt)= 0;
virtual bool getBottomLeftMapXY(orsPOINT2D * pt)= 0;
virtual bool getBottomRightMapXY(orsPOINT2D * pt)= 0;

//地理坐标
virtual bool getTopLeftLatLong(orsPOINT2D * pt)= 0;
virtual bool getTopRightLatLong(orsPOINT2D * pt)= 0;
virtual bool getBottomLeftLatLong(orsPOINT2D * pt)= 0;
virtual bool getBottomRightLatLong(orsPOINT2D * pt)= 0;
virtual bool getCenterLatLong(orsPOINT2D * pt)= 0;

public:
    ORS_INTERFACE_DEF(orsIMetaData, _T("band"))
};
```

4.3.6 元数据的读取

对于 XML 格式的元数据，先利用 getRDFService()—> readMetaDataFromXmlFile()从 XML 获取元数据的属性树，然后从属性树中读取各元数据的值。因为一般 XML 文件属于没有类型的字符串数据，一般需要把获取的元数据字符串值转换为需要的类型。为了便于使用，减少转换工作，对于整型、浮点等基本数值类型，OpenRS 会根据提取的类型自动尝试类型转换。如果转换失败，会返回 false。

4.3.7 太阳天顶角、方位角

大多数遥感探测器所接收的电磁辐射能量主要源自地球表面对太阳光的反射，而反射能量的大小和当时太阳与卫星星下点的相对位置有密切关系（谈小生等，1995）。一般可以根据影像元数据得到影像中心的太阳高度和方位信息。如果需要获得每个像元的太阳天顶角、方位角，那么可以根据像元的地理位置和成像的时间来计算。

这里采用中国气象科学研究院王炳忠研究员撰写的《太阳辐射计算讲座》(王炳忠, 1999)中的算法实现。具体代码参见 onsToaRadiance 插件中的 SunAngle.cpp。

4.3.8 观测天顶角、方位角

在卫星遥感所有参数里面,卫星天顶角和方位角是最复杂的。因为理论上,每个像元都不一样。如果知道卫星成像时的轨道信息和像元的地理位置,那么可以根据两者的连线计算出观测方位角和天顶角。

对于系统几何校正产品或正射影像产品,一般都不会提供成像时的轨道数据,而是直接提供观测天顶角和方位角的数据,以方便使用。如果提供每个像元的卫星角度数据,那么数据量就会很大。考虑到数据量的问题,一般只会提供一个抽稀的版本。以 HJ 卫星为例,HJ1A/1B 的 CCD 数据四个波段加起来 800 多兆(byte),如果每个像元都给出两个角度,那么角度数据也会有差不多 800 多兆。资源卫星中心提供下载的 level2 级别的 HJ CCD 数据里面的 SatAngle.txt 文件只有 10M,这是因为它是间隔采样的(33×33 个像素一个数据)。如果用户要用,那么需要自己进行插值处理。

OpenRS 的波段元数据接口 orsIBandMetaData 提供了一个传感器角度影像接口 getSensorAngleImg,用于获取与波段数据一致的观测角度数据。观测角度影像分为两个波段,波段 1 为观测天顶角,波段 2 为观测方位角。

4.4 简单要素矢量模块 orsSF

与栅格数据一样,矢量数据是遥感平台需要处理的关键数据之一。简单要素(simple feature)类型作为 OGC 的标准,得到了广泛的应用。我们基于 OGC 的标准定义了 OpenRS 的简单要素类型接口 osf(OpenRS simple feature)。osf 对象的根接口为 osfIObject。没有采用 orsIObject 的原因是,osf 对象作为矢量数据具有数据量大的特点,同时 osf 对象不需要 orsIObject 的属性、接口名称、对象描述等接口,可以适当简化。osfIObject 只保留了引用计数功能。

```
nterface osfIObject
{
    virtual void addRef()= 0;
    virtual void release()= 0;
};
```

osf 对象体系如图 4-10 所示,分为几何对象、属性对象、属性定义、属性域四大类。几何对象 osfIGeometry 又细分为点、线、多边形、几何集合等子对象,总体上与 OGC 标准一致。

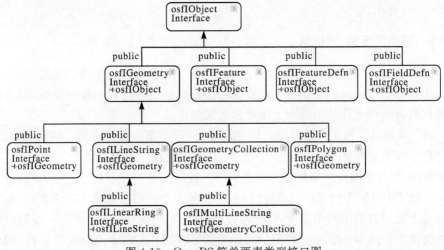

图 4-10 OpenRS 简单要素类型接口图

在实现上,OpenRS 目前内置了基于 GDAL OGR 模块的矢量数据读写功能,支持 ESRI Shapefile、ESRI ArcSDE、MapInfo (tab and mid/mif)、GML、KML、PostGIS、Oracle Spatial 等众多格式(详见 www.gdal.org/ogr)。

4.4.1 简单要素矢量数据源

简单要素矢量数据源接口 osfIVectorSource 定义了一个创建、打开、删除、管理简单要素数据源的方法。一个简单要素数据源是指一个简单要素文件、目录或数据库等多种形式的数据载体。在 OpenRS 中,最简单的简单要素数据源是一个 shape 文件或一个 shape 文件的目录。

```
struct osfIVectorSource : public orsIDataSource
{
public:
    //获取支持的格式
    virtual orsFileFormatList GetSupportedFormats()= 0;

    //创建指定格式和空间参考的数据源
    virtual bool Create(const char * pszName, char * pszFormat, orsISpatialRefer-
        ence * poSpatialRef)= 0;
    //打开给定的数据源
```

```cpp
    virtual bool Open(const char * pszName, int bUpdate)= 0;
    virtual bool IsOpen()const=0;
    //关闭该数据源
    virtual void Close()= 0;
    //删除给定的数据源
    virtual int DeleteDataSource(const char * pszName)= 0;

    //测试数据源的能力
    virtual int TestCapability(const char * )= 0;

    //获取图层的数目
    virtual int GetLayerCount()= 0;
    //获取给定的图层
    virtual osfIVectorLayer * GetLayer(int)= 0;

    //根据名字获取数据层
    virtual osfIVectorLayer * GetLayerByName(const char * )= 0;
    //删除给定的图层
    virtual OSFERR DeleteLayer(int)= 0;
    //创建一个给定类型和名字的图层
    virtual osfIVectorLayer * CreateLayer(const char * pszName,
        OSF_wkbGeometryType eGType, char * * papszOptions = NULL)= 0;

    ORS_INTERFACE_DEF(orsIDataSource, "SF");

};
```

4.4.2 简单要素矢量层

简单要素矢量层接口 osfIVectorLayer 定义了从简单要素矢量层读写简单要素的函数。

```cpp
struct osfIVectorLayer : public orsIObject
{
public:
    virtual osfIGeometry * GetSpatialFilter ()= 0;
    virtual void SetSpatialFilter(osfIGeometry * )= 0;
    virtual void SetSpatialFilterRect(double dfMinX, double dfMinY, double dfMaxX,
        double dfMaxY)= 0;
```

```cpp
    virtual osfIFeature * AppendFeature(OSF_wkbGeometryType type, orsPOINT3D * pts,
        int n) = 0;
    virtual bool AppendFeature(const osfIFeature * poFeature) = 0;
    virtual bool AppendFeatureDirectly(osfIFeature * poFeature) = 0;
    virtual OSFERR StoreFeature(long id) = 0;

    virtual void ResetReading() = 0;
    virtual osfIFeature * GetNextFeature() = 0;
    virtual OSFERR SetNextByIndex(long nIndex) = 0;
    virtual osfIFeature * GetFeature(long nFID) = 0;

    virtual OSFERR SetFeature(osfIFeature * poFeature) = 0;
    virtual OSFERR CreateFeature(osfIFeature * poFeature) = 0;
    virtual OSFERR DeleteFeature(long nFID) = 0;
    virtual OSFERR SyncToDisk() = 0;

    virtual osfIFeatureDefn * GetLayerDefn() = 0;
    virtual int GetFeatureCount(int bForce = true) = 0;
    virtual OSFERR GetExtent(osfEnvelope * psExtent, int bForce) = 0;

    virtual int TestCapability(const char *) = 0;
    virtual OSFERR CreateField(osfIFieldDefn * poField, int bApproxOK) = 0;
    virtual int TestDBF() = 0;
    virtual orsISpatialReference * GetSpatialReference() = 0;

public:
    ORS_INTERFACE_DEF(orsIObject, "osfIVectorLayer");
};
```

4.4.3 简单要素服务

简单要素服务 orsISFService 用于打开简单要素类型，实现几何对象的空间关系计算，创建简单要素对象。

```cpp
class orsISFService : public orsIService
{
public:
    virtual orsFileFormatList getSupposedSFFormats() = 0;
```

```cpp
virtual osfIVectorSource * OpenSFFile(const orsChar * filename, bool bUpdate) =
    0;
virtual osfIVectorSource * CreateDataSource(const orsChar * pszName, const ors-
    Char * format, orsISpatialReference * pSRS) = 0;
virtual int DeleteDataSource(const char * pszName) = 0;

//几何对象的空间计算
virtual int Intersects (osfIGeometry * pGeo, osfIGeometry * pGeoOther) = 0;
virtual int Equals (osfIGeometry * pGeo, osfIGeometry * pGeoOther) = 0;
virtual int Disjoint (osfIGeometry * pGeo, osfIGeometry * pGeoOther) = 0;
virtual int Touches (osfIGeometry * pGeo, osfIGeometry * pGeoOther) = 0;
virtual int Crosses (osfIGeometry * pGeo, osfIGeometry * pGeoOther) = 0;
virtual int Within (osfIGeometry * pGeo, osfIGeometry * pGeoOther) = 0;
virtual int Contains (osfIGeometry * pGeo, osfIGeometry * pGeoOther) = 0;
virtual int Overlaps (osfIGeometry * pGeo, osfIGeometry * pGeoOther) = 0;

//简单要素的创建
virtual osfIFeature * createFeature(osfIFeatureDefn * poDefnIn) = 0;
//简单要素属性域的定义
virtual osfIFieldDefn * createFieldDefn() = 0;
//几何对象的创建
virtual osfIGeometry * createGeometry(OSF_wkbGeometryType dataType) = 0;

//点对象的创建
virtual osfIPoint * createPoint() = 0;
//线对象的创建
virtual osfILineString * createLineString() = 0;
//环对象的创建
virtual osfILinearRing * createLinearRing() = 0;
//多边形对象的创建
virtual osfIPolygon * createPolygon() = 0;

//多点对象的创建
virtual osfIMultiPoint * createMultiPoint() = 0;
//多线对象的创建
virtual osfIMultiLineString * createMultiLineString() = 0;
//多个多边形对象的创建
```

```cpp
    virtual osfIMultiPolygon * createMultiPolygon()= 0;
    //几何对象集对象的创建
    virtual osfIGeometryCollection * createGeometryCollection()= 0;

    virtual bool startup(orsIPlatform * platform)=0;
    virtual void shutdown()=0;
    virtual bool isok()=0;

    ORS_INTERFACE_DEF(orsIService, "SF")
};
```

下面以简单要素数据源的打开为例说明服务的实现方式。OpenSFFile 用于打开给定数据源,bUpadate 表示是否要以修改的方式打开数据源。对于只读的情况,bUpadate 应为 false。OpenSFFile 首先查询已经注册的 osfIVectorSource 对象,然后逐个尝试这些对象是否能够打开给定的数据。

```cpp
osfIVectorSource * orsXSFService::OpenSFFile(const orsChar * filename, bool bUpdate)
{
    orsIRegisterService * registerService = getPlatform()->getRegisterService
        ();

    //根据接口名查询对象
    orsArray<ref_ptr<orsIObjectDesc>> descriptes =
        registerService->getObjectDescsByInterface("osfIVectorSource");
    …

    //逐个尝试
    for(int i=0;i< descriptes.size();i++)
    {
        orsIObjectDesc * desc = descriptes[i].get();

        //根据对象 ID 创建该对象
        osfIVectorSource * reader = ORS_CREATE_OBJECT(osfIVectorSource, desc->
            getID());

        //尝试打开
        if(reader != NULL && reader->Open(filename, bUpdate)){
            return reader;
```

        }
    }

    return NULL;
}

更多的使用方法,可以参考 6.3.5 节矢量层渲染器。

4.5 其他模块

4.5.1 基础地理数据管理模块 orsGeoData

在遥感影像处理中,经常要使用全球或区域 DEM 或 DOM 数据。SRTM、ASTER、ETM 全球数据是典型的全球基础地理数据。为了简化这些数据的使用,作者设计开发了 orsGeoData 模块。

orsGeoData 模块实现全球 DEM、DOM 数据的动态拼接与读写。getDEMFileName、getDOMFileName 根据给定的期望的分辨率,自动在索引中查询与给定影像 pRangeImg 重叠的影像,并自动形成虚拟拼接文件名。GetDEM、GetDOM 与 getDEMFileName、getDOMFileName 功能一致,只是返回的是已经打开的栅格数据接口。

```
interface orsIGeoDataService : public orsIService
{
public:
    //取 DEM 文件名,一个时为原始文件名,多个时为文件名列表文件
    virtual orsString getDEMFileName(float desiredGSD, const orsIImageSource *
        pRangeImg, int fromRow, int toRow, float *retGSD)= 0;
    virtual orsIImageSource * GetDEM(float desiredGSD, const orsIImageSource *
        pRangeImg, int fromRow, int toRow, float *retGSD)= 0;

    //取 DOM 文件名,一个时为原始文件名,多个时为文件名列表文件
    //对于控制点可以用这个返回控制点列表
    virtual orsString getDOMFileName(float desiredGSD, const orsIImageSource *
        pRangeImg, int fromRow, int toRow, float *retGSD)= 0;

    virtual orsIImageSource * GetDOM(float desiredGSD, const orsIImageSource *
        pRangeImg, int fromRow, int toRow, float *retGSD)= 0;
```

```
public:
    ORS_INTERFACE_DEF(orsIService, "geoData")
};
```

为了实现全球地理数据的查询,设计了二级索引。第一级以分辨率为主索引,第二级以影像文件的经纬度范围为索引。第一级索引存在 etc\geoData 目录下。第二级索引文件 orsGeoData.idx 存于各影像目录下。文件的格式为 ASCII 码。第一行为文件标识"OpenRS GeoData Index",第二行为文件数,第三行开始为经纬度-文件索引。示例文件如下。

```
OpenRS GeoData Index
15
111.000000 33.999722 112.000278 35.000000    N30E110/ASTGTM_N34E111_dem.tif
112.000000 33.999722 113.000278 35.000000    N30E110/ASTGTM_N34E112_dem.tif
113.000000 33.999722 114.000278 35.000000    N30E110/ASTGTM_N34E113_dem.tif
```

1. 一级索引文件

在 etc\geoData 目录下,indexedDEMs.txt 和 indexedDOMs.txt 保存了 DEM 和 DOM 的一级编目信息。以 indexedDEMs.txt 为例,文件第一行为二级索引树,第二行开始为分辨率-文件目录索引。

```
2
   90.0 O:\GeoData\SRTM_90m\
 1000.0 O:\GeoData\SRTM_1km\
```

2. 实用程序

1) 编目程序 GeoDataCatalog

GeoDataCatalog.exe 是一个命令行程序,参数为待编目的影像目录。GeoDataCatalog 将自动遍历所有子目录,在给定目录下创建索引文件 orsGeoData.idx(图 4-11)。

2) 地理数据自动提取程序 GeoDataExtractor

GeoDataExtractor 是一个命令行程序,用法如下:

```
Usage: GeoDataExtractor DEM/DOM rangRasterFile outputFile requireGsd
```

GeoDataExtractor 根据指定的是 DEM 还是 DOM,从编目数据中提取与

第 4 章 面向遥感影像处理的基础模块

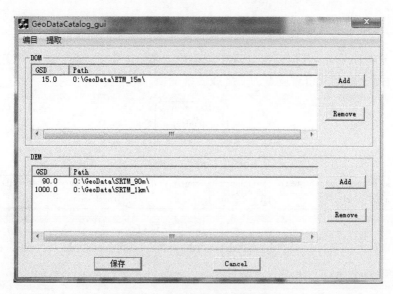

图 4-11 编目与索引管理界面

rangRasterFile 相交的满足要求分辨率 requireGsd 的拼接数据文件 outputFile（图 4-12）。

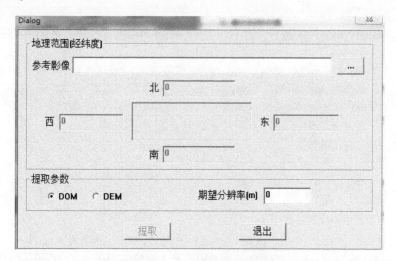

图 4-12 地理数据提取对话框

3) 编目与索引管理界面程序 GeoDataCatalog_gui

这是一个编目与索引管理界面程序，用于调用 GeoDataCatalog 和 GeoDataExtractor，管理以及一级索引。"编目"菜单自动调用编目程序，实现对给定目录的索引和"提取"菜单。

4.5.2 几何变换处理模块 orsGeometry

如图 4-13 所示,几何变换处理模块提供二维、三维、二维-三维几何变换的解算功能,用于确定两组对应点的二维或三维变换。

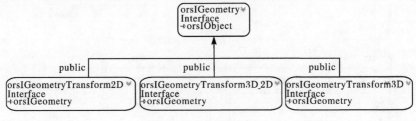

图 4-13 几何变换接口

1. 二维变换

建立两组二维点的二维仿射变换、相似变换等二维变换。

```
interface orsIGeometryTransform2D: public orsIGeometry
{
public:
    virtual void Initialize(const orsPOINT2D * ptsSrc, int n, const orsPOINT2D *
        ptsDst, float * weights = NULL)= 0;
    virtual void GetResidual(const orsPOINT2D * ptsSrc, int n, const orsPOINT2D *
        ptsDst, orsPOINT2D * pVxys)= 0;
    virtual void GetMeanError(double * mx, double * my)= 0;

    //中心化的参数
    virtual void GetParameter(orsPOINT2D * pcSrc, orsPOINT2D * pcDst, double * a,
        double * b = NULL)= 0;

    //不需中心化的参数
    virtual void GetParameter(double * a, double * b = NULL)= 0;
    virtual void Transform(const orsPOINT2D * ptsSrc, int n, orsPOINT2D * ptsDst)
        = 0;
    virtual void Transform(double * x, double * y)= 0;

    ORS_INTERFACE_DEF(orsIGeometry, "transform2D");
};
```

2. 三维变换

建立两组三维点的之间的三维仿射变换、相似变换等三维变换。

```
interface orsIGeometryTransform3D: public orsIGeometry
{
public:
    virtual void Initialize(const orsPOINT3D * ptsSrc, int n, const orsPOINT3D *
        ptsDst, float * weights = NULL) = 0;
    virtual void GetResidual(const orsPOINT3D * ptsSrc, int n, const orsPOINT3D *
        ptsDst, orsPOINT3D * pVxys) = 0;
    virtual void GetMeanError(double * mx, double * my, double * mz) = 0;

    //中心化的参数
    virtual void GetParameter(orsPOINT3D * pcSrc, orsPOINT3D * pcDst, double * a,
        double * b, double * c) = 0;

    //不需中心化的参数
    virtual void GetParameter(double * a, double * b, double * c) = 0;
    virtual void Transform(const orsPOINT3D * ptsSrc, int n, orsPOINT3D * ptsDst)
        = 0;

    ORS_INTERFACE_DEF(orsIGeometry, "transform3D");
};
```

3. 三维至二维的变换

实现三维坐标至二维坐标的投影变换,如直接线性变换。

```
interface orsIGeometryTransform3D_2D: public orsIGeometry
{
public:
    virtual void Initialize(const orsPOINT3D * ptsSrc, int n, const orsPOINT2D *
        ptsDst, float * weights = NULL) = 0;
    virtual void GetResidual(const orsPOINT3D * ptsSrc, int n, const orsPOINT2D *
        ptsDst, orsPOINT2D * pVxys) = 0;
    virtual void GetMeanError(double * mx, double * my) = 0;

    //中心化的参数
```

```
    virtual void GetParameter(orsPOINT3D * pcSrc, orsPOINT2D * pcDst, double * a,
        double * b = NULL,  double * c = NULL)= 0;

//不需中心化的参数
    virtual void GetParameter(double * a, double * b = NULL, double * c = NULL)
        = 0;
    virtual void Transform(const orsPOINT3D * ptsSrc, int n, orsPOINT2D * ptsDst)
        = 0;

    ORS_INTERFACE_DEF(orsIGeometry, "transform3D_2D");
};
```

第5章 界面扩展设计

界面扩展是 OpenRS 的重要设计目标之一。期望在插件动态库中定义的菜单、工具条、窗口等界面对象能够动态插入宿主程序的界面中,以增强 OpenRS 的处理能力。本章首先介绍 MFC 扩展库 BCGControlBar,然后基于 BCGControl-Bar 的扩展机制介绍 OpenRS 界面扩展的设计与实现。

5.1 BCGControlBar 简介

BCGControlBar 是一个基于 MFC 的扩展库(汪雷,2014),如图 5-1 所示,可以通过完全的用户化操作构成一些类似于 Microsoft Office 2000/XP/2003 和 Microsoft Visual Studio.NET 的应用程序(用户工具栏、菜单、键盘等)。BCGControlBar 库包含了 150 多个经过精心设计、测试和具有完备文档的 MFC 扩展类。这些都可以很容易地应用于应用程序,节省大量的开发和调试时间。BCGControlBar 专业版的扩展库包含大量高级界面,如可分离的表窗口,自动隐藏窗体,拖拽时能够显示停靠控件和工具栏的内容,新增停靠算法(类似于在 Microsoft Visual Studio.NET 环境和 Microsoft Visio 中引入的算法),具有可分离的快捷栏、制表工具栏、语法检验和 In-telliSense-style 支持的文本控件,完全的平面视觉等。除了与 MFC 兼容的控制栏的执行部分,这个库的其他结构完全是由 BCG 公司自己设计的。大部分新的系统特性都是由库的内核自动激活和管理的。

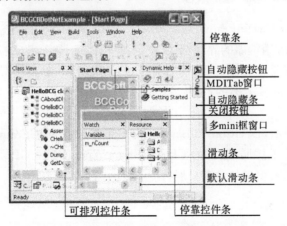

图 5-1 BCGControlBar 的典型界面元素

5.1.1 选择 BCGControlBar 的理由

1. 为什么不选择 Codejock 公司的 Xtreme

在选择 BCGControlBar 还是 Codejock 公司的 Xtreme 的问题上,作者纠结了很久。外观上,Xtreme 可能在某些方面比 BCGControlBar 要美观,但微软在更新 MFC 时选择了俄罗斯的 BCGControlBar 而不是美国的 Xtreme。微软 msdn 的博客上写道:"…An important consideration in working with BCGSoft was that their architecture made it easy to integrate into the existing MFC library…"(http://blogs.msdn.com/b/vcblog/archive/2008/04/07/mfc-update-powered-by-bcgsoft.aspx)。

通过研究发现,BCGControlBar 更适合于按插件的方式进行界面扩展。基于 BCGControlBar 的界面扩展显得很自然。而 Xtreme 的实现机制在某种程度上破坏了 MFC 的架构。例如,在 Mainframe 中 Xtreme 采用回调函数 OnDocking-PaneNotify、OnCreateCommandBar、OnCreateControl 三个函数来实现界面的扩展,这对于基于插件的界面动态扩展编程显得极不自然。而在 BCGControlBar 中只需要按照 MFC 的方式在 OnCreate 中直接创建扩展对象就可以了。

2. 为什么不直接使用更新后的 MFC

MFC 更新是从 Visual C++ 2008 开始的,作者希望 OpenRS 能支持包括 Visual C++ 6.0 在内的所有 Visual C++ 版本,这一点 BCGControlBar 做得很好。

5.1.2 BCGControlBar 的扩展性

在 MFC 中,CMainFrame 类是 MFC 为应用程序的主框架窗口创建的框架派生类,它定义了应用程序的界面特性,包括菜单栏、工具栏、状态栏和窗口等。

若应用程序是单文档的(称为 SDI),则 CMainFrame 类从 CFrameWnd 类派生;若应用程序是多文档的(称为 MDI),则 CMainFrame 类从 CMDIFrameWnd 类派生。

如图 5-2 所示,BCGControlBar 在 MFC 的基础上对 CFrameWnd 和 CMDIFrameWnd 进行了扩展,定义的 CBCGPFrameWnd 和 CBCGPMDIFrameWnd 都支持菜单、工具条、窗口的动态扩展。

1. 菜单的动态插入

CBCGPMenuBar 是 BCGControlBar 的菜单条对象,在 MFC 菜单对象的基础

上实现了菜单条的停靠和"最近使用菜单"等功能。CBCGPMenuBar 继承自 CB-CGPToolBar，菜单的插入功能通过继承 CBCGPToolBar 的 InsertButton 功能实现。CBCGPToolBar::InsertButton 的函数原型如下：

```
int InsertButton(const CBCGPToolbarButton&button, INT_PTR iInsertAt=-1);
int InsertButton(CBCGPToolbarButton * pButton, int iInsertAt=-1);
```

功能是在当前工具条的 iInsertAt 位置插入新的按钮 pButton。其中，pButton 是菜单按钮或下拉菜单。

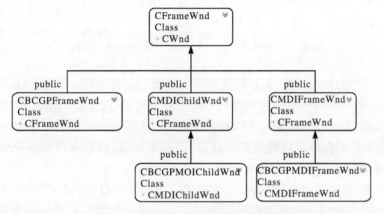

图 5-2　BCGControlBar 重载的框架窗口类

2. 工具条的动态创建

BCGControlBar 的工具条 CBCGPToolBar 支持停靠等多种功能，可以直接调用 CBCGPToolBar 的 Create 或 CreateEx 创建。

```
BOOL Create(CWnd * pParentWnd, DWORD dwStyle, UINT nID);
BOOL CreateEx(CWnd * pParentWnd, DWORD dwCtrlStyle, DWORD dwStyle, CRect rcBorders,
    UINT nID);
```

其中，pParentWnd 是要在其中创建工具条的框架对象指针，为 CBCGP-FrameWnd、CBCGPMDIFrameWnd 类；nID 为资源 ID。

3. 窗口控件的创建

BCGControlBar 的 CBCGPDockingControlBar 是支持停靠的控件条，支持一般的控件和窗口。可以直接在 CBCGPFrameWnd、CBCGPMDIFrameWnd 调用 CBCGPControlBar 的 Create 或 CreateEx 创建。

```
BOOL Create(LPCTSTR lpszCaption, CWnd * pParentWnd, const RECT&rect, BOOL bHasGrip-
    per, UINT nID, DWORD dwStyle, DWORD dwTabbedStyle, DWORD dwBCGStyle, CCreateCon-
    text * pContext);
BOOL CreateEx ( DWORD dwStyleEx, LPCTSTR lpszCaption, CWnd * pParentWnd, const
    RECT&rect, BOOL bHasGripper, UINT nID, DWORD dwStyle, DWORD dwTabbedStyle, DWORD
    dwBCGStyle, CCreateContext * pContext);
```

其中,pParentWnd 是要在其中创建工具条的框架对象指针,为 CBCGP-FrameWnd、CBCGPMDIFrameWnd 类;nID 为资源 ID。

5.2 OpenRS 界面扩展接口与实现

OpenRS 的插件对象界面采用 MFC 和 BCGControlPro 实现。MFC 实现界面的框架,BCGControlPro 实现界面布局的动态定制(图 5-3)。OpenRS 界面扩展的设计思路是定义抽象框架接口 orsIFrameWnd 和框架扩展接口 orsIGuiExtension。框架扩展插件实现框架扩展接口,而抽象框架接口 orsIFrameWnd 在宿主程序中通过模板 orsIFrameWndHelper 实现。

在加载实现 orsIGuiExtension 的插件对象时,主程序把抽象框架接口 orsIFrameWnd 指针传递给框架扩展插件对象。插件对象就可以通过抽象框架接口 orsIFrameWnd 向宿主程序的主框架添加菜单、工具条和控件窗口(图 5-4)。

这里的 orsIGuiExtension 类似于 Eclipse 的扩展点,就是指可以通过插件在此处动态增加功能。

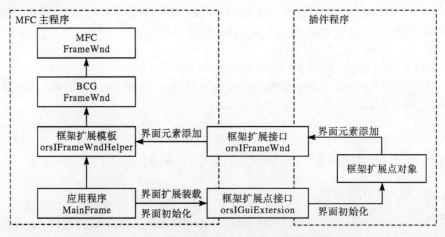

图 5-3 以 MFC 框架和 BCG 为基础的界面的扩展机制

第 5 章　界面扩展设计

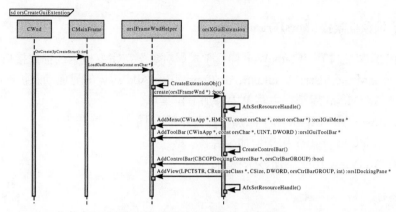

图 5-4　界面扩展对象创建顺序图

5.2.1　框架扩展接口 orsIGuiExtension

框架扩展接口 orsIGuiExtension 是框架扩展对象必须实现的接口，宿主程序通过框架扩展点接口查询所有实现该接口的框架扩展对象，然后创建这些对象并加载。

```
#include "orsBase/orsIExtension.h"

interface orsIGuiExtension : public orsIExtension
{
public:
    virtual bool create(orsIFrameWnd * frameWnd) = 0;
    virtual bool pluginMenu(orsIFrameWnd * frameWnd) = 0;

    virtual BOOL OnCmdMsg(UINT nID, int nCode, void * pExtra, AFX_CMDHANDLERINFO *
        pHandlerInfo) = 0;
    virtual LRESULT windowProc ( HWND hWnd, UINT message, WPARAM wParam, LPARAM
        lParam) = 0;

public:
    ORS_INTERFACE_DEF(orsIExtension, "gui")
};
```

OpenRS 界面扩展对象接口 orsIGuiExtension 定义了界面扩展对象的创建方式和消息处理方式。函数 create 用于界面扩展对象的创建。扩展对象 create 函数中创建菜单、工具条、窗口等界面元素，并通过 frameWnd 添加到宿主程序框架中。

5.2.2 抽象框架接口 orsIFrameWnd

抽象框架接口 orsIFrameWnd 向框架扩展对象暴露宿主框架的抽象接口。该接口包括 AddMenu、InsertMenuItem、AddControlBar、AddView 等界面动态扩展函数。

（1）AddMenu 用于在主菜单中添加下拉菜单。

（2）InsertMenuItem 用于添加下拉菜单项（目前版本屏蔽了）。

（3）AddControlBar 用于增加控件条。

（4）AddView 用于添加显示视图。

```
interface orsIFrameWnd
{
    virtual CWnd * GetCWnd() = 0;

    //增加下拉菜单, menuRcId 为菜单条菜单. menuStr 为挂在主菜单上的名字,若为 NULL,
    则为带名字的多级菜单
    virtual orsIGuiMenu * AddMenu(CWinApp * pWinApp, UINT menuRcId, const orsChar
        * menuStr, const orsChar * beforeWhich = NULL) = 0;
    //增加下拉菜单, menuRcId 为菜单条菜单. menuStr 为挂在主菜单上的名字,若为 NULL,
    则为带名字的多级菜单
    virtual orsIGuiMenu * AddMenu(CWinApp * pWinApp, HMENU hMenu, const orsChar *
        menuStr, const orsChar * beforeWhich = NULL) = 0;
    //增加下拉菜单项,如在 View 下增加显示、隐藏工具条、视图的菜单项
    virtual bool InsertMenuItem(CWinApp * pWinApp, const orsChar * which, const ors-
        Char * insertAfter, UINT menuID) = 0;


    //增加工具条
    virtual orsIGuiToolBar * AddToolBar(CWinApp * pWinApp, const orsChar * lpszWin-
        dowName, UINT toolBarRcId, DWORD dwAlignment = CBRS_ALIGN_ANY) = 0;
    //增加控制条
    virtual bool AddControlBar(CBCGPDockingControlBar * pControlBar, orsCtrlBar-
        GROUP group = ORS_CTRLBARGROUP_NONE) = 0;


    //增加视图
    virtual orsIDockingPane * AddView(LPCTSTR lpszWindowName, CRuntimeClass * pRun-
        TimeClass, CSize sizeDefault, DWORD dwStyle, orsCtrlBarGROUP group = ORS_
        CTRLBARGROUP_NONE) = 0;


    virtual orsIDockingPane * AddView2(LPCTSTR lpszWindowName, CRuntimeClass *
```

```
    pRunTimeClass, CSize sizeDefault, DWORD dwStyle, orsCtrlBarGROUP group, int
    uid=0)=0;

virtual void ShowControlBar(CBCGPBaseControlBar * pBar, BOOL bShow, BOOL bDe-
    lay, BOOL bActivate) = 0;

//分配 Command ID
virtual intAllocCmdID() = 0;
};
```

AddControlBar 和 AddView 的 orsCtrlBarGROUP 为加入的控件或视图的组。目前预定义了类似于 Visual Studio 的工作空间组 WORKSPACE 和结果输出组 OUTPUT。枚举类型定义如下：

```
enum orsCtrlBarGROUP{
    ORS_CTRLBARGROUP_NONE,
    ORS_CTRLBARGROUP_WORKSPACE,
    ORS_CTRLBARGROUP_OUTPUT
};
```

5.2.3 OpenRS 界面元素

OpenRS 界面元素包括菜单、工具条、可停靠的窗口、视图等。其中，可停靠的窗口和视图都抽象为可停靠框，对应于 BCGControlBar 的 BCGPDockingControlBar。

1. OpenRS 界面元素接口

OpenRS 界面元素包括菜单、工具条和停靠框（图 5-5）。

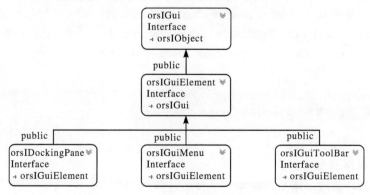

图 5-5　OpenRS 界面元素接口

1) 界面对象接口

本接口只起到分类的作用。

```
interface orsIGui: public orsIObject
{
    ORS_INTERFACE_DEF(orsIObject, "gui");
};
```

2) 界面元素对象接口

界面元素对象接口定义隐藏和显示两个函数。

```
interface orsIGuiElement : public orsIGui
{
public:
    virtual bool show() = 0;
    virtual bool hide() = 0;


public:
    ORS_INTERFACE_DEF(orsIGui, "element");
};
```

3) 菜单对象接口

菜单对象接口用于持有一个菜单对象。

```
interface orsIGuiMenu : public orsIGuiElement
{
public:
    virtual bool Init(CBCGPMenuBar * pMenuBar, CWinApp * pWinApp, HMENU hMenu, const
        orsChar * menuStr, const orsChar * beforeWhich) = 0;
    virtual bool Init(CBCGPMenuBar * pMenuBar, CWinApp * pWinApp, UINT menuRcId,
        const orsChar * menuStr, const orsChar * beforeWhich) = 0;

public:
    ORS_INTERFACE_DEF(orsIGuiElement, "menu");
};
```

4) 工具条对象接口

工具条对象接口用于持有一个工具条对象。

```cpp
interface orsIGuiToolBar : public orsIGuiElement
{
public:
    virtual bool Init(orsIFrameWnd * pFrmWnd, CBCGPToolBar * theToolBar) = 0;

public:
    ORS_INTERFACE_DEF(orsIGuiElement, "toolbar");
};
```

5) 停靠框对象接口

停靠框对象接口用于持有一个可停靠对象。

```cpp
class CBCGPDockingControlBar;
interface orsIDockingPane : public orsIGuiElement
{
public:
    virtual BOOL Create(LPCTSTR lpszWindowName, CWnd * pParentWnd,
        CSize sizeDefault, BOOL bHasGripper, UINT nID, DWORD dwStyle =
        WS_CHILD | WS_VISIBLE | CBRS_TOP | CBRS_HIDE_INPLACE, DWORD dwTabbedStyle =
        CBRS_BCGP_REGULAR_TABS, DWORD dwBCGStyle = dwDefaultBCGDockingBarStyle)
        = 0;

public:
    virtual CWnd * AttachView(CRuntimeClass * pViewClass) = 0;
    virtual CWnd * GetWnd() = 0;
    virtual void Detach() = 0;
    virtual CBCGPDockingControlBar * GetControlBar() = 0;
    virtual void EnableContextMenu(bool bEnable = true) = 0;

public:
    ORS_INTERFACE_DEF(orsIGuiElement, "dockingPane");
};
```

2. OpenRS 界面元素实现

1) 菜单对象实现

```cpp
class orsXGuiMenu : public orsIGuiMenu, orsObjectBase
{
```

```cpp
private:
    CMenu m_menu;//MFC 菜单对象
    CBCGPMenuBar * m_pMenuBar;//BCG 菜单条对象
    const orsChar * m_menuStr;//下拉菜单名
    const orsChar * m_beforeWhich;//菜单插入位置

    int m_nShowCount;

public:
    orsXGuiMenu()
    {
        m_nShowCount = 0;
        m_pMenuBar = NULL;
        m_beforeWhich = NULL;
    }

    virtual ~orsXGuiMenu();
    //初始化函数,输入的是菜单句柄 hMenu
    virtual bool Init(CBCGPMenuBar * pMenuBar, CWinApp * pWinApp, HMENU hMenu, const
        orsChar * menuStr, const orsChar * beforeWhich)
    {
        m_nShowCount = 0;
        m_pMenuBar = pMenuBar;//菜单条对象
        m_menuStr = menuStr;//菜单名
        m_beforeWhich = beforeWhich;//插入位置
        m_menu.Attach(hMenu);//菜单句柄关联
        //菜单来自插件?如果是,则对菜单项的资源 ID 进行和谐处理
        if(pWinApp)
            getGuiResIDService()->HarmonyResourceID(pWinApp, hMenu);
        //显示菜单
        show();

        return true;
    }
    //初始化函数,输入的是菜单资源 ID
    virtual bool Init(CBCGPMenuBar * pMenuBar, CWinApp * pWinApp, UINT menuRcId,
        const orsChar * menuStr, const orsChar * beforeWhich)
    {
```

```
    m_nShowCount = 0;
    m_pMenuBar = pMenuBar;
    //装载菜单
    m_menuStr = menuStr;
    m_beforeWhich = beforeWhich;
    m_menu.LoadMenu(menuRcId);

    if(pWinApp)
        getGuiResIDService()->HarmonyResourceID(pWinApp, m_menu.GetSafeHmenu());

    show();

    return true;
}
//菜单显示
virtual bool show()
{
    if(m_nShowCount == 0){
        int insertAt = -1;
        if(NULL != m_beforeWhich){
            int n;

            HMENU hMenu = m_pMenuBar->GetHMenu();
            //菜单项数
            n = ::GetMenuItemCount(hMenu);

            orsChar menuStr[80];
            for(int i=0; i< n ; i++)
            {
                //取菜单项名
                GetMenuString(hMenu, i, menuStr, 80, MF_BYPOSITION);
                //是否是插入位置?
                if(NULL != strstr(menuStr, m_beforeWhich)){
                    insertAt = i;
                    break;
                }
            }
        }
}
```

```
        if(NULL == m_menuStr){
            CString menuString;
            m_menu.GetMenuString(0, menuString, MF_BYPOSITION);

            m_pMenuBar->InsertButton(
                CBCGPToolbarMenuButton(0, m_menu.GetSubMenu(0)->GetSafeHmenu
                (), -1, menuString),
                insertAt);
        }
        else{
            const orsChar * localString = NULL;

            //本地化,取本地化菜单名
            if(getGuiLocalizationService())
                localString =getGuiLocalizationService()->GetLocalizedString
                    (m_menuStr);
            //找到本地化名字,用本地化名字插入菜单按钮
            if(localString)
                m_pMenuBar->InsertButton(
                    CBCGPToolbarMenuButton(0, m_menu.GetSafeHmenu(), -1,
                    localString),insertAt);
            else
                //用原始名字插入菜单按钮
                    m_pMenuBar->InsertButton(
                        CBCGPToolbarMenuButton(0, m_menu.GetSafeHmenu(), -1, m_
                        menuStr),insertAt);
        }
    }
    //调整布局
    m_pMenuBar->AdjustLayout();
    m_pMenuBar->AdjustSizeImmediate();
}
m_nShowCount++;
return true;
}

virtual bool hide()
```

```cpp
    {
        --m_nShowCount;
        return true;
    }

public:
    ORS_OBJECT_IMP1(orsXGuiMenu, orsIGuiMenu, _T("BCG"), "GUI Menu");
};
```

2) 工具条对象实现

```cpp
class orsXGuiToolBar : public orsIGuiToolBar, orsObjectBase
{
private:
    CBCGPToolBar *m_pToolBar;//BCG工具条件指针
    orsIFrameWnd *m_pFrmWnd;//框架指针
    int m_nShowCount;

public:
    orsXGuiToolBar()
    {
        m_nShowCount = 0;
        m_pToolBar = NULL;
    }
    virtual ~orsXGuiToolBar()
    {
        if(m_pToolBar)
            delete m_pToolBar;
    }
    //初始化,记录框架指针和工具条指针
    virtual bool Init(orsIFrameWnd *pFrmWnd, CBCGPToolBar *theToolBar)
    {
        m_pFrmWnd = pFrmWnd;
        m_pToolBar = theToolBar;
        return true;
    }
    //工具条显示
    virtual bool show()
    {
```

```
        //如果未显示,则在框架中显示工具条
        if(!m_pToolBar->IsWindowVisible())
            m_pFrmWnd->ShowControlBar(m_pToolBar, TRUE, FALSE, TRUE);

        return true;
        }
    //隐藏工具条
    virtual bool hide()
    {
        //如果工具条已显示,则在框架中隐藏工具条
        if(m_pToolBar->IsWindowVisible())
            m_pFrmWnd->ShowControlBar(m_pToolBar, FALSE, FALSE, FALSE);

        return true;
    }

public:
    ORS_OBJECT_IMPl(orsXGuiToolBar, orsIGuiToolBar, _T("BCG"), "GUI Menu");
};
```

3) 停靠框对象实现

```
#include "orsBCGPDockingPane.h"
class orsXDockingPane : public orsIDockingPane, orsObjectBase
{
public:
    orsXDockingPane()
    {
        m_pContainedWnd = NULL;
    }

    virtual ~orsXDockingPane()
    {
    }

    virtual CWnd * GetWnd()
    {
        return m_pContainedWnd;
```

```cpp
    };

    virtual CWnd * AttachView(CRuntimeClass * pViewClass)
    {
        //绑定视图
        m_pContainedWnd = m_dockingPane.AttachView(pViewClass);

        return m_pContainedWnd;
    }

    virtual void Detach()
    {
        //解绑视图
        m_dockingPane.Detach();
    }

    virtual BOOL Create(LPCTSTR lpszWindowName, CWnd * pParentWnd, CSize sizeDefault,
        BOOL bHasGripper, UINT nID, DWORD dwStyle, DWORD dwTabbedStyle, DWORD dwBCGStyle)
    {
        //创建停靠框
        return m_dockingPane.Create(lpszWindowName, pParentWnd, sizeDefault,
            bHasGripper, nID, dwStyle, dwTabbedStyle, dwBCGStyle);
    }

    virtual bool show()
    {
        //显示停靠框
        return m_dockingPane.ShowWindow(SW_SHOW);
    }

    virtual bool hide()
    {
        //隐藏停靠框
        return m_dockingPane.ShowWindow(SW_HIDE);
    }

    virtual void EnableContextMenu(bool bEnable = true)
    {
```

```
        //允许或关闭上下文菜单
        m_dockingPane.EnableContextMenu(bEnable);
    }
    //取 BCG 停靠框指针

    virtual CBCGPDockingControlBar *GetControlBar()
    {
        return &m_dockingPane;
    }

private:
    CWnd *m_pContainedWnd;

    orsBCGPDockingPane m_dockingPane;

public:
    ORS_OBJECT_IMP2(orsXDockingPane,  orsIDockingPane, orsIGuiElement,
        "default", "BCG");
};
```

其中，orsBCGPDockingPane 是 CBCGPDockingControlBar 的扩展。

```
class orsBCGPDockingPane : public CBCGPDockingControlBar
{
private:
    CWnd *m_pWndChild;
    //绑定的视图类指针
    CRuntimeClass *m_pViewClass;
    bool m_bContextMenuAllowed;

public:
    orsBCGPDockingPane();
    {
        m_pWndChild = NULL;
        m_pViewClass = NULL;

        m_bContextMenuAllowed = true;
    }
    virtual ~orsBCGPDockingPane();
```

```cpp
    void EnableContextMenu(bool bEnable = true){
        m_bContextMenuAllowed = bEnable;
    };

public:
    //绑定视图类
    CWnd * AttachView(CRuntimeClass * pViewClass)
    {
        m_pViewClass = pViewClass;
        //创建视图
        CWnd * pView = (CWnd *)pViewClass->CreateObject();
        if(pView == NULL)
        {
            TRACE1("Warning: Dynamic create of view type %hs failed.\n",
                pViewClass->m_lpszClassName);
            return NULL;
        }

        ASSERT_KINDOF(CWnd, pView);
        //创建的视图窗口句柄
        m_pWndChild = pView;

        return pView;
    }


    //调用该方法解除 pane 和用户窗口的关系
    void Detach()
    {
        if(m_pWndChild)
        {
            m_pWndChild->ShowWindow(SW_HIDE);
            m_pWndChild->SetParent(NULL);
            m_pWndChild = NULL;
        }
    }


protected:
```

```cpp
//{{AFX_MSG(CWorkspaceBar)
afx_msg int OnCreate(LPCREATESTRUCT lpCreateStruct);
afx_msg void OnSize(UINT nType, int cx, int cy);
afx_msg void OnClose();
afx_msg void OnDestroy();
afx_msg void OnContextMenu(CWnd * pWnd, CPoint point);
//}}AFX_MSG
DECLARE_MESSAGE_MAP()
};
```

在orsBCGPDockingPane.cpp中:

```cpp
int orsBCGPDockingPane::OnCreate(LPCREATESTRUCT lpCreateStruct)
{
    if(CBCGPDockingControlBar::OnCreate(lpCreateStruct)==-1)
        return -1;
    if(m_pWndChild == NULL)
        return -1;

    if(!m_pWndChild->Create(NULL, NULL, WS_CHILD | WS_CLIPCHILDREN | WS_CLIPSIBLINGS,
        CRect(0,0,0,0), this, 0, NULL))
    {
        TRACE0("Warning: could not create view for frame.\n");
        return NULL;          //创建视图失败,返回NULL
    }

    m_pWndChild->ShowWindow(SW_NORMAL);

    return 0;
}
void orsBCGPDockingPane::OnSize(UINT nType, int cx, int cy)
{
    CBCGPDockingControlBar::OnSize(nType, cx, cy);

    if(m_pWndChild)
    {
        m_pWndChild->SetWindowPos(NULL, -1, -1, cx, cy, SWP_NOMOVE |
            SWP_NOZORDER | SWP_NOACTIVATE);
```

```
            }
    }

    void orsBCGPDockingPane::OnClose()
    {
        CBCGPDockingControlBar::OnClose();
    }


    void orsBCGPDockingPane::OnDestroy()
    {
        CBCGPDockingControlBar::OnDestroy();
    }
    //取上下文菜单
    void orsBCGPDockingPane::OnContextMenu(CWnd * pWnd, CPoint point)
    {
        if(m_bContextMenuAllowed)
            CBCGPControlBar::OnContextMenu(pWnd, point);
    }
```

5.2.4 框架扩展实现模板 orsIFrameWndHelper

OpenRS 通过一个帮助性的模板类 orsIFrameWndHelper 来实现 CBCGP-FrameWnd 和 CBCGPMDIFrameWnd 的扩展接口。主要功能包括菜单扩展、工具条扩展、窗口扩展、视图的扩展。其中,视图的扩展是窗口扩展的特例。

1. orsIFrameWndHelper 的数据成员

为了能够实现不同界面元素的插件化处理,在 orsIFrameWndHelper 中定义了界面空间、菜单、工具条、界面扩展对象等多种界面对象集合。

```
//工作空间组,用于包含要添加到工作组的控制条
CBCGPDockingControlBar * m_pWorkSpaceTabbedBar;
//输出窗口组,用于包含要添加到输出窗口组的控制条
CBCGPDockingControlBar * m_pOutputTabbedBar;
//菜单集合
std::map<UINT, ref_ptr<orsIGuiMenu> >m_mapMenus;
//工具条集合
std::map<UINT, ref_ptr<orsIGuiToolBar> >m_mapToolBars;
//停靠控件集合
```

```cpp
std::vector<CBCGPDockingControlBar *> m_vCtrlBars;
//界面扩展对象
orsArray<ref_ptr<orsIGuiExtension> >m_vExtensions;
```

2. 动态菜单添加函数

增加下拉菜单，hMenu 为菜单条菜单。menuStr 为挂在主菜单上的名字，如果为 NULL，则为带名字的多级菜单

```cpp
virtual orsIGuiMenu * AddMenu(CWinApp * pWinApp, HMENU hMenu, const orsChar * menuStr, const orsChar * beforeWhich = NULL)
{
    //调用界面服务创建一个菜单对象
    orsIGuiMenu *pGuiMenu = m_guiService->CreateMenu();
    //在主菜单条 m_wndMenuBar 中添加菜单 hMenu
    pGuiMenu->Init(&m_wndMenuBar, pWinApp, hMenu, menuStr, beforeWhich);

    //记录该菜单对象
    m_mapMenus.insert(std::make_pair(UINT(hMenu), pGuiMenu));

    return pGuiMenu;
}
```

增加下拉菜单，menuRcId 为菜单条菜单 ID。menuStr 为挂在主菜单上的名字，如果为 NULL，则为带名字的多级菜单。

```cpp
virtual orsIGuiMenu * AddMenu(CWinApp * pWinApp, UINT menuRcId, const orsChar * menuStr, const orsChar * beforeWhich = NULL)
{
    //查看菜单是否已经创建过
    _resid2MenuMapIter iter = m_mapMenus.find(UINT(pWinApp+menuRcId));
    if(iter != m_mapMenus.end())
        return iter->second.get();

    //调用界面服务创建一个菜单对象
    orsIGuiMenu *pGuiMenu = m_guiService->CreateMenu();
    //初始化菜单
    pGuiMenu->Init(&m_wndMenuBar, pWinApp, menuRcId, menuStr, beforeWhich);
```

```cpp
    //记录该菜单对象
    m_mapMenus.insert(std::make_pair(UINT(pWinApp+menuRcId), pGuiMenu));

    return pGuiMenu;
}
```

3. 动态工具条添加函数

```cpp
virtual orsIGuiToolBar * AddToolBar(CWinApp * pWinApp, const orsChar * lpszWindowN-
    ame, UINT toolBarRcId, DWORD dwAlignment)
{
    //查看工具条是否已经创建过
    _resid2ToolBarMapIter iter = m_mapToolBars.find(UINT(pWinApp+toolBarRcId));

    if(iter != m_mapToolBars.end())
        return iter->second.get();
    //创建菜单
    CBCGPToolBar * theToolBar = new CBCGPToolBar;
    if(!theToolBar->CreateEx(this, TBSTYLE_FLAT, WS_CHILD | WS_VISIBLE |
        CBRS_GRIPPER|CBRS_TOOLTIPS|CBRS_FLYBY | CBRS_SIZE_DYNAMIC, CRect(1,1,1,
        1),m_nextToolBarID)|| !theToolBar->LoadToolBar(toolBarRcId))
    {
        TRACE0("Failed to create toolbar\n");
        return false;
    };

    //如果是插件动态库对象增加的工具条,则修改工具条菜单项的命令ID,以保持唯
      一性
    if(pWinApp)
        m_guiResIdService->HarmonyResourceID(pWinApp, theToolBar);
    //工具条 ID 加 1
    m_nextToolBarID++;

    theToolBar->SetWindowText(lpszWindowName);
    theToolBar->EnableDocking(dwAlignment);
    //停靠该工具条
```

```
    DockControlBar(theToolBar);

    //创建工具条对象
    orsIGuiToolBar *pGuiToolBar = m_guiService->CreateToolBar();
    pGuiToolBar->Init(this, theToolBar);
    //记录工具条对象
    m_mapToolBars.insert(std::make_pair(UINT(pWinApp+toolBarRcId), pGuiTool-
        Bar));

    return pGuiToolBar;
}
```

4. 动态控件条添加函数

```
virtual bool AddControlBar(CBCGPDockingControlBar *pControlBar, orsCtrlBarGROUP
    group)
{
    //记录工具条对象
    m_vCtrlBars.push_back(pControlBar);

    //加入指定的组空间
    switch(group){
    case ORS_CTRLBARGROUP_WORKSPACE:
    pControlBar->EnableDocking(CBRS_ALIGN_ANY);
    DockControlBar(pControlBar);

    if(m_pWorkSpaceTabbedBar)
        pControlBar->AttachToTabWnd(m_pWorkSpaceTabbedBar, BCGP_DM_SHOW, TRUE,
            &m_pWorkSpaceTabbedBar);
    else
        m_pWorkSpaceTabbedBar = pControlBar;
    break;
    case ORS_CTRLBARGROUP_OUTPUT:
        pControlBar->EnableDocking(CBRS_ALIGN_BOTTOM);
        DockControlBar(pControlBar);

        if(m_pOutputTabbedBar)
```

```
            pControlBar->AttachToTabWnd(m_pOutputTabbedBar, BCGP_DM_SHOW, TRUE,
                &m_pOutputTabbedBar);
        else
            m_pOutputTabbedBar = pControlBar;
        break;
    default:
        pControlBar->EnableDocking(CBRS_ALIGN_ANY);
        DockControlBar(pControlBar);
    }
    return false;
}
```

5. 动态视图添加函数

```
//增加视图
virtual orsIDockingPane * AddView(LPCTSTR lpszWindowName, CRuntimeClass * pRunTime-
    Class, CSize sizeDefault, DWORD dwStyle, orsCtrlBarGROUP group)
{
    ASSERT(m_guiService);
    if(m_guiService){
        //创建停靠窗格
        orsIDockingPane * thePane = m_guiService->CreatDockingPane();
        //根据运行时类添加视图
        thePane->AttachView(pRunTimeClass);

        //创建该窗格
        if(!thePane->Create(lpszWindowName, this,sizeDefault,
            TRUE, AllocCmdID(), dwStyle))
        {
            TRACE0("Failed to create Class View bar\n");
            return false;        //fail to create
        }

        //工具条ID加1
        m_nextToolBarID++;
        AddControlBar(thePane->GetControlBar(), group);
```

```
        return thePane;
    }
    return NULL;
}
```

5.2.5 消息处理与 ID 和谐

MFC 资源 ID 是一个整型变量,在一个运行的程序进程中必须具有唯一性。所以,如果需要在 MFC 的动态库中定义菜单和事件,必须预先规定好不同动态库中 ID 的范围。对于基于插件的编程,这是几乎不可能实现的。为了克服这一困难,OpenRS 中专门设计和定义了界面资源 ID 服务,用于协调不同动态库的 ID,保证正确事件响应。如图 5-6 所示,消息处理在传到插件的 WinApp 对象之前,先经过界面服务对象 orsIGuiResIDSeverce 进行 ID 映射处理,把进程中的唯一 ID 映射回插件编译时的 ID。

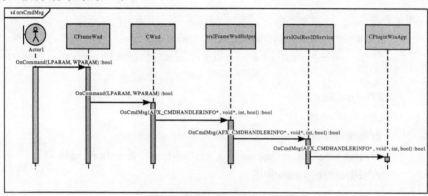

图 5-6　OpenRS 界面消息处理

1. 界面资源 ID 服务接口 orsIGuiResIDService

界面资源 ID 服务包含动态 ID 获取接口、菜单及工具条 ID 和谐接口、事件消息分发接口。其中,动态 ID 获取接口用于获取下一个唯一的 ID 号;菜单及工具条 ID 和谐接口用于自动替换已经加载的菜单和工具条的 ID;事件消息分发接口用于把和谐后的消息转为各插件动态库的原始资源 ID 并分发到相应动态库。

```
interface orsIGuiResIDService : public orsIService
{
public:
    //可用的下一个命令 ID 设置
```

```cpp
    virtual void SetNextCmdID(UINT nextCmdID) = 0;
    //获取下一个可用命令 ID
    virtual UINT GetNextCmdID() = 0;

public:
    //和谐模块的资源 ID
    virtual bool HarmonyResourceID(CWinApp * pWinApp, HMENU hMenu) = 0;
     virtual bool HarmonyResourceID(CWinApp * pWinApp, CBCGPToolBar * pToolBar)
        = 0;

    //取和谐后 ID 的所有者和原始 ID,如果是主程序 ID,则返回 NULL
    virtual CWinApp * GetIDOwner(UINT uniqueId, UINT * oldId) = 0;

    //把和谐后的命令分发到原模块进行下一步的处理
    virtual bool OnCmdMsg(UINT nID, int nCode, void * pExtra, AFX_CMDHANDLERINFO
        * pHandlerInfo) = 0;

    ...
};
```

2. 界面资源 ID 服务实现

```cpp
//新旧 ID 映射表
struct uniqueID
{
    newID newId;
    oldID oldId;
};

//模块内 ID
struct appID {
    CWinApp * app;
    oldID oldId;
};

//新 ID 到模块内 ID 的映射表
typedef std::map<newID, appID>newID_appIdMap;
//模块内 ID 数据表(一个模块有多个 ID)
```

```
typedef std::multimap<CWinApp *, uniqueID>app_uniqIDMap;
//系统 ID 到响应模块表(一个系统命令可能有多个响应模块)
typedef std::multimap<sysID, CWinApp *>sysID_appMap;
```

1) 唯一 ID 的获取

```
//获取进程内唯一、插件(DLL)内部唯一对应的 ID
UINT orsXGuiResIDService::getUniqueID(CWinApp * pWinApp, UINT oldId)
{
    //如果是系统命令,则不需转换
    if(IsSystemCommand(oldId)){
        //加入系统命令响应表,取模块包含 ID 的范围
        sysID_appMapIter low = m_mapSysID2App.lower_bound(oldId);
        sysID_appMapIter upper = m_mapSysID2App.upper_bound(oldId);
        while(low != upper)
        {
            //该 ID 是否已存在
            if(low->second == pWinApp){
                return 0;
            }
            ++low;
        }
        //生成并插入新的映射项
        m_mapSysID2App.insert(std::make_pair(oldId, pWinApp));

        return 0;
    }

    //取模块包含 ID 的范围
    app_uniqIDMapIter low = m_mapApp2UniqIDs.lower_bound(pWinApp);
    app_uniqIDMapIter upper = m_mapApp2UniqIDs.upper_bound(pWinApp);

    while(low != upper)
    {
        //直接返回,app 已经登记
        if(low->second.oldId == oldId){
            return low->second.newId;
        }
```

```cpp
        ++low;
    }

    //分配新的 ID
    UINT newId = m_nextCmdID++;

    //记录新 ID
    {
        appID appId;

        appId.app = pWinApp;
        appId.oldId = oldId;
        m_mapNewId2AppID[newId] = appId;

        uniqueID uniId;
        uniId.oldId = oldId;
        uniId.newId = newId;
        m_mapApp2UniqIDs.insert(std::make_pair(pWinApp, uniId));
    }

    return newId;
}
```

2) 工具条 ID 的和谐

```cpp
bool orsXGuiResIDService::HarmonyResourceID(CWinApp * pWinApp, CBCGPToolBar *
    pToolBar)
{
    ASSERT(NULL != pToolBar);

    //如果 pWinApp == NULL,那么说明是主模块
    if(NULL == pWinApp || NULL == pToolBar)
        return false;
    //取工具条按钮个数
    int nItem = pToolBar->GetCount();
    //工具条按钮循环
    for(int i=0; i<nItem; i++)
    {
        //取唯一 ID
```

```cpp
    UINT newId = getUniqueID(pWinApp, pToolBar->GetItemID(i));

    //替换 ID
    if(newId > 0)
        pToolBar->GetButton(i)->m_nID = newId;
    }

    return true;
}
```

3) 弹出式菜单 ID 的和谐

```cpp
bool orsXGuiResIDService::HarmonyPopupMenuID(CWinApp * pWinApp, HMENU hMenu)
{
    int nItem = ::GetMenuItemCount(hMenu);
    UINT state;
    //菜单项循环
    for(int i=0; i<nItem; i++)
    {
        state = ::GetMenuState(hMenu, i, MF_BYPOSITION);

        if(state & MF_POPUP){//子菜单,和谐子菜单中的字符串
            //本地化
            if(getGuiLocalizationService()){
                //临时变量
                TCHAR sMenuAddString[128] = {0};

                ::GetMenuString(hMenu, i, sMenuAddString, 128, MF_BYPOSITION);

                const orsChar * localString = NULL;
                //取本地化字符串
                localString = getGuiLocalizationService()->GetLocalizedString
                    (sMenuAddString);
                //替换菜单中的字符串
                    if(localString)
                        ::ModifyMenu(hMenu, i, MF_STRING|MF_BYPOSITION, 0,
                            localString);
            }
```

```
            //和谐子菜单 ID(递归)
            HarmonyPopupMenuID(pWinApp, ::GetSubMenu(hMenu, i));
        }

        else if(state & MF_SEPARATOR)//分隔条
            continue;
        else
        //和谐菜单项
        {
            UINT oldId = ::GetMenuItemID(hMenu, i);
            UINT newId = getUniqueID(pWinApp, oldId);

            if(newId > 0){
                TCHAR sMenuAddString[128] = {0};

                ::GetMenuString(hMenu, oldId, sMenuAddString, 128, MF_BYCOMMAND);

                const orsChar * localString = NULL;

                //本地化
                if(getGuiLocalizationService())
                    localString = getGuiLocalizationService()->GetLocalized-
                        String(sMenuAddString);

                if(localString)
                    ::ModifyMenu(hMenu, oldId, MF_STRING|MF_BYCOMMAND, newId,
                        localString);

                else
                    ::ModifyMenu(hMenu, oldId, MF_STRING|MF_BYCOMMAND, newId,
                        sMenuAddString);
            }
        }
    }

    return true;
}
```

4) 菜单 ID 的和谐

```cpp
//协调模块的资源 ID
bool orsXGuiResIDService::HarmonyResourceID(CWinApp * pWinApp, HMENU hMenu)
{
    int nItem = ::GetMenuItemCount(hMenu);
    UINT state;

    for(int i=0; i<nItem; i++)
    {
        state = ::GetMenuState(hMenu, i, MF_BYPOSITION);

        if(state & MF_POPUP){//子菜单
            //本地化
            if(getGuiLocalizationService()){
                TCHAR sMenuAddString[128] = {0};
                    ::GetMenuString(hMenu, i, sMenuAddString, 128, MF_BYPOSITION);

                const orsChar * localString = NULL;
                localString = getGuiLocalizationService()->GetLocalizedString
                    (sMenuAddString);
                if(localString)
                    ::ModifyMenu(hMenu, i, MF_STRING|MF_BYPOSITION, 0, localString);
            }
            HarmonyResourceID(pWinApp, ::GetSubMenu(hMenu, i));
        }

        else if(state & MF_SEPARATOR)//分隔条
            continue;
        else
        {
            UINT oldId = ::GetMenuItemID(hMenu, i);
            UINT newId = getUniqueID(pWinApp, oldId);

            if(newId > 0){
                TCHAR sMenuAddString[128] = {0};
                ::GetMenuString(hMenu, oldId, sMenuAddString, 128, MF_BYCOMMAND);
```

```cpp
        const orsChar * localString = NULL;
        //本地化
        if(getGuiLocalizationService())
            localString = getGuiLocalizationService()->GetLocalized-
                String(sMenuAddString);

        if(localString)
            ::ModifyMenu(hMenu, oldId, MF_STRING|MF_BYCOMMAND, newId, lo-
                calString);
        else
            ::ModifyMenu(hMenu, oldId, MF_STRING|MF_BYCOMMAND, newId,
                sMenuAddString);
        }
    }
}

    return 0;
}
```

5) 取唯一 ID 的所有者

```cpp
//取协调后 ID 的所有者和原始 ID,如果是公有的则返回 NULL
CWinApp * orsXGuiResIDService::GetIDOwner(UINT uniqueID, UINT * oldID)
{
    *oldID = uniqueID;
    //查找 ID 映射表
    newID_appIdMapIter found = m_mapNewId2AppID.find(uniqueID);
    //未找到
    if(found == m_mapNewId2AppID.end())
        return NULL;

    *oldID = found->second.oldId;
    //返回模块指针
    return found->second.app;
}
```

6) 命令消息的处理

```cpp
bool orsXGuiResIDService::OnCmdMsg(UINT nID, int nCode, void * pExtra, AFX_CMDHAN-
```

```cpp
    DLER-INFO * pHandlerInfo)
{
    //不是命令消息?
    if(nCode != CN_UPDATE_COMMAND_UI && nCode != CN_COMMAND)
        return false;

    UINT oldId;
    //取 ID 的模块和旧 ID
    CWinApp * pWinApp = GetIDOwner(nID, &oldId);
    //若 ID 来自插件,则转发消息到插件模块进行处理
    if(pWinApp)
        return pWinApp->OnCmdMsg(oldId, nCode, pExtra, pHandlerInfo);
    //若是系统命令,则逐个调用所有模块处理
    if(IsSystemCommand(nID)){
        sysID_appMapIter low = m_mapSysID2App.lower_bound(nID);
        sysID_appMapIter upper = m_mapSysID2App.upper_bound(nID);
        if(low != upper){
            while(low != upper)
            {
                low->second->OnCmdMsg(nID,nCode,pExtra,pHandlerInfo);
            }
            return true;
        }
    }

    return false;
}
```

5.2.6 插件中的界面对象创建与消息处理

下面结合 orsViewerExt_vecEdit 来说明插件界面的创建与消息处理。如图 5-7 所示,orsViewerExt_vecEdit 是 orvViewer 的一个用于矢量编辑的扩展插件(参见 6.4.3 节),带有一个下拉菜单和两个工具条。界面元素的创建在界面扩展对象的 create 函数中进行,界面元素的消息转发到界面扩展对象的 OnCmdMsg 进行处理。

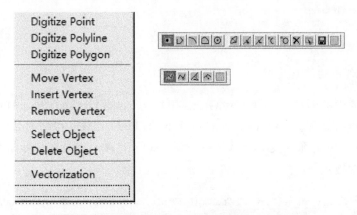

图 5-7 orsViewerExt_vecEdit 的菜单和工具条

1. 插件中的界面元素创建

主程序的框架创建时，会调用插件的 create 函数供创建自己的菜单、工具条和控件窗口。在创建自己的界面元素前后，要保存现场和恢复现场。

```
bool orsViewerExt_vecEdit::create(orsIFrameWnd * frameWnd)
{
    //保存资源环境
    HINSTANCE oldRcHandle = AfxGetResourceHandle();
    //切换到本插件的资源
    AfxSetResourceHandle(GetModuleHandle("orsViewerExt_vecEdit.dll"));
    //保存框架窗口接口
    m_pParentFrameWnd = frameWnd;

    //在框架中增加菜单
    frameWnd->AddMenu(&g_theApp, IDR_VECTOR_MENU, _T("Vector"), _T("帮助"));
    //在框架中增加工具条
    orsIGuiToolBar * pDigBar = frameWnd->AddToolBar(&g_theApp, _T("Digitize
        Bar"), DR_DIGITIZER_BAR, CBRS_ALIGN_ANY);
    //在框架中增加工具条
    frameWnd->AddToolBar(&g_theApp, _T("Snap Bar"), IDR_SNAP_BAR);

    //恢复资源环境
    AfxSetResourceHandle(oldRcHandle);
```

```
    return true;
}
```

2. 插件中的消息处理

插件界面的消息经界面 orsXGuiResIDService 的 OnCmdMsg 解析后分派到相应插件进行处理。orsViewerExt_vecEdit 的 OnCmdMsg 接到消息后转发给插件的 CWinApp 的派生类 COrsViewerExt_vecEdit_App 进行进一步解析，以调用各消息的处理函数。

```
BOOL orsViewerExt_vecEdit::OnCmdMsg(UINT nID, int nCode, void *pExtra, AFX_CMDHAN-
    DLERINFO *pHandlerInfo)
{
    if(g_theApp.OnCmdMsg(nID, nCode, pExtra, pHandlerInfo))
    return TRUE;

    return FALSE;
}
```

其中，g_theApp 为 COrsViewerExt_vecEdit_App 的单例对象。

COrsViewerExt_vecEdit_App 的 OnCmdMsg 先绕过 App 有效性检查，然后调用 CWinApp::OnCmdMsg 进行真正的处理。

```
COrsViewerExt_vectEdit_App g_theApp;

BOOL COrsViewerExt_vecEdit_App::OnCmdMsg(UINT nID, int nCode, void *pExtra, AFX_
    CMDHANDLERINFO *pHandlerInfo)
{
    //绕过 App 有效性检查
    AFX_MANAGE_STATE(AfxGetStaticModuleState());
    pCtlState = (myAFX_MAINTAIN_STATE2 *)&_ctlState;
    if(TRUE == CWinApp::OnCmdMsg(nID, nCode, pExtra, pHandlerInfo))
        return TRUE;

    return FALSE;
}
```

本节介绍了 orsViewerExt_vecEdit 的创建和消息处理，只涉及菜单和工具条，包含界面空间的界面插件请参考 orsViewerExt_workflow。

5.3 属 性 界 面

OpenRS 的一个设计目标是为插件对象的运行提供一个统一的参数配置界面。以可执行对象为例,可以把可执行对象的参数分为输入文件、输出文件、运行参数、处理范围四类。为插件对象提供统一的界面,也是实现界面无关的算法对象,使算法与界面分离,从而实现桌面处理、分布式并行处理一体化的前提。

本节首先介绍 BCG 的属性控件机制及内置属性,然后针对遥感数据处理的需要,对 BCG 的属性机制进行扩展,以满足面向插件的处理框架的需要。

5.3.1 BCG 属性

1. BCG 属性控件

CBCGPProp 是 BCGControlBar 的重要组成部分,实现了一系列具有编辑功能的属性控件。每一个 BCG 的属性可以表示特定的数据类型。BCG 的属性可以包含子属性,构成一个树状属性界面。这一点正好和 OpenRS 的属性树对应,如果能使两者有机结合,则有可能实现 OpenRS 属性的统一编辑界面。

目前,BCG 内置的属性控件类型如表 5-1 所示。

表 5-1 BCG 内置属性控件类型

属性名称	中文含义
String	字符串
Numeric(with the spin button support)	数值
Real	实数
List of options(combo box)	选项列表
Boolean	布尔变量
Color	颜色列表
File/directory	文件或目录
Font	字体选择
Date/Time(Professional Edition only)	日期/时间
Brush	填充刷
Line style	线型
Text format	文本格式

2. BCG 属性的类别

从属性编辑的角度看,BCG 的属性可以分为三类,即简单变量类、就地编辑类和对话框类。简单变量类包括字符串、实数、布尔变量等;就地编辑类包括选项列表、颜色列表、日期列表等。就地编辑类属性框被单击时会在最右侧出现下拉箭头,并向下弹出一个选项框进行属性值的选择,如图 5-8 所示。

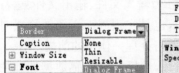 

图 5-8　就地编辑类属性

和就地编辑类不同,在单击属性值时,对话框类属性框会在最右侧出现两个点的符号,单击该符号会弹出一个对话框,进行属性的编辑操作。如图 5-9～图 5-11 所示,典型的对话框类属性包括文件或目录、字体等。

图 5-9　目录对话框属性

图 5-10　文件框属性

除了内置的属性类型，BCG 支持属性控件类型的扩充，最基本的就是自定义对话框属性(图 5-12)。

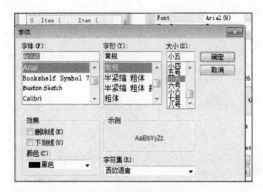

图 5-11　字体对话框属性

图 5-12　自定义对话框属性

3. BCG 属性列表

BCG 的属性 CBCGPProp 由属性列表 CBCGPPropList 进行管理。一般先创建一个属性列表，并与对话框界面控件关联，然后在属性列表中添加属性项即可实现属性的显示与编辑。示例代码如下：

```
CBCGPPropList m_wndPropList;

CBCGPProp * pGroup1 = new CBCGPProp(_T("Group"));
pGroup1->AddSubItem(new CBCGPProp(_T("Item 1"),(_variant_t)false, _T("Descrip-
    tion")));
pGroup1->AddSubItem(new CBCGPProp(_T("Item 2"),(_variant_t)2501, _T("Descrip-
    tion")));

m_wndPropList.AddProperty(pGroup1);
```

CBCGPPropList 的主要功能包括属性添加、删除、属性值变化回调、命令设置等。

1）添加属性

```
int AddProperty(CBCGPProp * pProp,BOOL bRedraw=TRUE,BOOL bAdjustLayout=TRUE);
```

本函数在属性列表的尾部添加新的属性。

2）删除属性

```
BOOL DeleteProperty(CBCGPProp * &pProp, BOOL bRedraw = TRUE, BOOL bAdjustLayout =
    TRUE);
```

本函数删除指定的属性及其子属性。

3）删除所有属性

```
void RemoveAll();
```

本函数删除属性表的所有属性。

4）通过用户自定义数据寻找对应属性项

```
CBCGPProp * FindItemByData(DWORD_PTR dwData,BOOL bSearchSubItems=TRUE)const;
```

通过用户定义的 ID 查找属性。该属性值是属性创建时，通过 SetData 函数设置的。

5）添加下拉选项

```
BOOL AddOption(LPCTSTR lpszOption,BOOL bInsertUnique=TRUE);
```

AddOption 添加下拉选项到下拉框，供用户选择。

6）删除所有下拉选项

```
void RemoveAllOptions();
```

RemoveAllOptions 删除所有下拉选项。

7）添加子属性

```
BOOL AddSubItem(CBCGPProp * pProp);
```

AddSubItem 把子属性添加到现有属性形成多级属性。

8) 删除子属性

```
BOOL RemoveSubItem(CBCGPProp * &pProp, BOOL bDelete=TRUE);
```

RemoveSubItem 把子属性从当前属性中删除。

9) 恢复原值

```
virtual void ResetOriginalValue();
```

在编辑一个属性前，BCG 会保留原始属性值。ResetOriginalValue 用于恢复到原始值。

10) 设置用户定义数据

```
void SetData(DWORD_PTR dwData);
```

为当前属性保存一个用户自定义的长整数值。

11) 取回用户定义数据

```
DWORD_PTR GetData()const;
```

取回一个用户自定义的长整数值。

12) 属性值改变回调

```
virtual void OnPropertyChanged(CBCGPProp * pProp)const;
```

属性值变化时该函数被 BCGControlBar 框架调用，可以被派生类重载。其中，pProp 是值发生变化的属性对象。

属性值变化的回调机制是实现动态属性界面的基础。

5.3.2 OpenRS 自定义属性界面

尽管 BCG 提供了数值、下拉选项、颜色、字体、线型、文件等多种常用的属性，但是并不能完全满足遥感软件的开发需要。例如，在遥感数据处理中，经常会遇到以下问题：

（1）从已打开的影像图层列表中选择影像进行处理。
（2）从已打开的影像中选择一个或多个波段进行处理。
（3）定义影像纠正或其他处理结果的空间参考系统。
（4）选择一组影像进行拼接裁剪。
（5）切换不同的算法进行处理，而不同的算法可能具有不同的参数。
……

因为不同的处理对象需要的属性不同,难以预先进行完整的定义,因而无法预先做好相应的属性界面。作者希望 OpenRS 的属性界面和 OpenRS 的插件一样也是能够动态扩展的。

1. OpenRS 自定义属性描述

为了实现属性界面和算法对象的分离,并且保证属性界面的简单易用,作者期望充分利用 OpenRS 的资源描述的便利性(参见 3.4.1 节 RDF 服务),通过属性的语义描述来配置属性的界面。

首先,从属性描述的角度,作者定义了 EnumString、ClassID、CustomProp、PropDlg 等关键字,通过这些关键字来识别一个属性是否为自定义属性。其中,CustomProp 对应 BCG 的就地编辑类、PropDlg 对应对话框类属性。

1) 枚举类型属性

EnumString 表示一个枚举类型。例如,

〈SampleMethod〉〈IsA〉〈EnumString〉

〈SampleMethod〉〈Desc〉〈Sample Method String〉

〈SampleMethod〉〈Enum〉〈Nearest〉

〈SampleMethod〉〈Enum〉〈Bilinear〉

〈SampleMethod〉〈Enum〉〈BiCubic〉

表示 SampleMethod 是一个枚举字符串,可以是 Nearest、Bilinear、BiCubic 等采样方法。在创建属性界面时将自动把枚举的采样方法添加到下拉列表中供选择(图 5-13)。

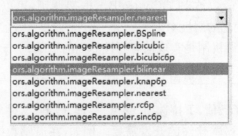

图 5-13 采样方法属性界面

2) 算法 ID 类型属性

ClassID 表示该属性是一个某一类对象的 ID,例如,

〈ResampleAlgID〉〈IsA〉〈ClassID〉

〈ResampleAlgID〉〈Desc〉〈ClassID of Resampler〉

〈ResampleAlgID〉〈ClassInterface〉〈orsIAlgResampler〉

表示 ResampleAlgID 是对象 ID，对象的接口是 orsIAlgResampler。在创建 ResampleAlgID 的属性界面时会把 orsIAlgResampler 所有对象的 ID 添加到下拉列表中供选择(图 5-14)。

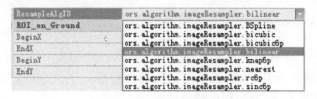

图 5-14　采样算法属性界面

3) 文件名属性

定义 FileName 为自定义属性，并且具有文件扩展名和文件类型属性。

⟨FileName⟩⟨IsA⟩⟨CustomProp⟩

⟨FileName⟩⟨NameOf⟩⟨File⟩

⟨FileName⟩⟨Has⟩⟨FileExt⟩

⟨FileName⟩⟨Has⟩⟨FileType⟩

根据 FileName 的自定义属性和文件类型，会在创建属性对话框时进行专门处理，如自动从图层树提取文件列表等。

⟨ImageFileName⟩⟨IsA⟩⟨FileName⟩

⟨ImageFileName⟩⟨FileType⟩⟨Image⟩

⟨ImageFileName⟩⟨FileExt⟩⟨Image TIFF(*.tif)|*.tif|ERDAS(*.img)|*.img|ENVI
　　(*.hdr)|*.hdr|All Files(*.*)|*.*||⟩

⟨MSImageFileName⟩⟨IsA⟩⟨ImageFileName⟩

这样，通过 RDF 推理 MSImageFileName 属于 FileName 的属性，也将成为一个自定义属性。

图 5-15 中，多光谱影像的属性为影像文件名，在创建时会自动把已打开的全色影像和多光谱影像作为下拉选项添加到该属性界面。

Property	Value	
□ 输入文件名		
多光谱影像	G:\data\fusion\02NOV21044347-M2AS_R1C2-000000185940_01_P001.TIF	...
全色影像	G:\data\fusion\02NOV21044347-M2AS_R1C2-000000185940_01_P001.TIF	
□ 输出文件名		
融合影像	$(MS_ImageSource)_Brovey.tif	

图 5-15　影像文件名属性界面示例

4) 自定义对话框属性

PropDlg 表示一个自定义属性具有自己的对话框，例如，

⟨ColorTable⟩⟨IsA⟩⟨CustomProp⟩
⟨ColorTable⟩⟨PropDlg⟩⟨ors.extension.propDlg.ColorTableConfigure⟩

表示 ColorTable 具有属性对话框 ors.extension.propDlg.ColorTableConfigure（图 5-16）。创建该属性时，将根据给定的属性创建一个关联的对话框对象的按钮，当单击该按钮时，会触发相应的对话框。

图 5-16　色表属性界面

2. OpenRS 自定义属性界面设计

针对 BCG 的就地编辑属性和对话框属性类，作者定义了 orsIPropControl 和 orsIPropDlg 两种对应的属性界面接口（图 5-17），而 orsIPropControlList 对应于 BCG 的属性列表 CBCGPPropList。

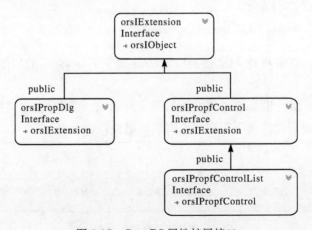

图 5-17　OpenRS 属性扩展接口

图 5-18 为部分 OpenRS 属性控件，基于 orsIPropControl 接口作者实现了列表属性、文件名属性、波段号属性、多光谱波段集属性和自定义对话框属性。由该图可以看出，orsIPropControl 对应于 BCG 的 CBCGPProp。orsIPropControl 具有与 CBCGPProp 类似的函数，但不依赖于 BCG，目的在于定义形式上的属性界面，而与具体的实现分离。

1）OpenRS 属性控件接口 orsIPropControl

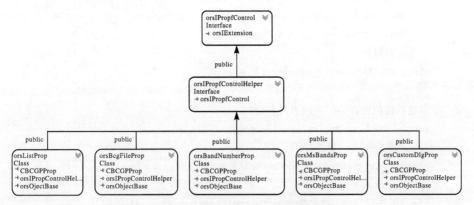

图 5-18　部分 OpenRS 属性控件

orsIPropControl 的接口定义如下，主要提供属性的绑定、子属性的添加和删除等动态属性界面实现需要的功能。

```
interface orsIPropControl : public orsIExtension
{
public:
    //取属性名
    virtual const orsChar * getPropName() = 0;

    //设置父属性控件
    virtual void setParent(orsIPropControl * pParent) = 0;
    //取父属性控件
    virtual orsIPropControl * getParent() = 0;

    //添加子属性控件
    virtual void addSubPropControl(orsIProperty * pPropGroup, const orsChar
        * name) = 0;
    //添加子属性控件
    virtual void addSubPropControl(orsIPropControl * pControl) = 0;
```

```cpp
    //删除子属性控件
    virtual void removeSubPropControl(const orsChar *name) = 0;
    //查找子属性控件
    virtual orsIPropControl *findPropControl(const orsChar *propName) = 0;

    //绑定 OpenRS 属性
    virtual void bindOrsProperty(orsIProperty *pPropGroup, const orsChar *prop-
        Name, orsVariantType propType) = 0;
    //取 BCG 属性
    virtual CBCGPProp *getBcgProp() = 0;

    //把属性值放在临时属性表内并返回
    virtual bool getTempData(orsIProperty *) = 0;
    //更新数据到 Bcg 属性界面
    virtual void updateData2BcgControl() = 0;
    //从 Bcg 取回数据
    virtual void updateDataFromBcgControl() = 0;

    virtual void Enable() = 0;
    virtual void Disable() = 0;

public:
    ORS_INTERFACE_DEF(orsIExtension, "propControl")
};
```

2) OpenRS 属性列表接口 orsIPropControlList

orsIPropControlList 除具有 orsIPropControl 的功能,可以增删子属性外,还可以设置属性事件监听器,用于响应属性的变化。

```cpp
interface orsIPropControlList : public orsIPropControl
{
public:
    virtual void addPropControl(orsIPropControl *pControl) = 0;
    virtual void removeAll() = 0;
    virtual void setListener(orsPropertyListener *listener) = 0;
public:
    ORS_INTERFACE_DEF(orsIPropControl, "list")
};
```

orsPropertyListener 是一个基于委托模型的事件监听器,可以作为 OpenRS 对象的一个特殊属性加到属性表中。

```
using namespace fastdelegate;
typedef FastDelegate2<orsIPropControlList *, orsIPropControl *, bool> orsProper-
    tyListener;

#define ORS_PROPERTY_LISTNER _T("PropertyListner")
```

3) OpenRS 属性对话框接口 orsIPropDlg

orsIPropDlg 实际是和 orsCustomDlgProp 配合使用的。在 orsCustomDlg-Prop 控件框最右侧的双点符号被单击时,会调用 orsIPropDlg 弹出实际对话框进行属性编辑。

```
interface orsIPropDlg : public orsIExtension
{
public:
    //从 prop 提取 attrName 属性,设置完成后返回
    virtual bool runDlg(orsIProperty * prop, const orsChar * attrName, orsIProp-
        Control * pParent) = 0;
public:
    ORS_INTERFACE_DEF(orsIExtension, "propDlg")
};
```

目前 OpenRS 已实现的对话框属性包括文件列表、水平坐标系统等(图 5-19)。

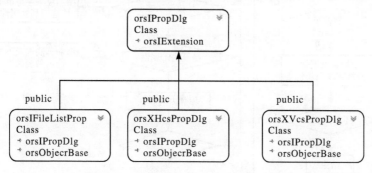

图 5-19 已实现的 OpenRS 属性对话框

3. OpenRS 属性列表的创建过程

为了统一地创建属性界面,作者专门在 orsIGuiService 中定义了 AddBCG-PropertyTree 函数,用于属性列表的创建。

```
virtual void AddBCGPropertyTree(orsIPropControlList * wndPropList,
    orsIExecute * pExeObj,orsILayerCollection * pLayerCollection = NULL);
```

AddBCGPropertyTree 可以自动地把一个可执行对象的属性转化为 BCG 的属性树,递归地创建属性组和属性(图 5-20)。首先,AddBCGPropertyTree 会根据需要在属性列表 wndPropList 创建输入文件名、输出文件名、运行参数和处理范围等标准属性组;然后,会调用子属性添加过程自动添加一个属性组内的属性。在添加属性过程中,会根据 RDF 推理,判断一个属性是否属于自定义类型、自定义对话框类型、文件名类型、枚举类型、算法 ID 类型等创建特定的属性对象。

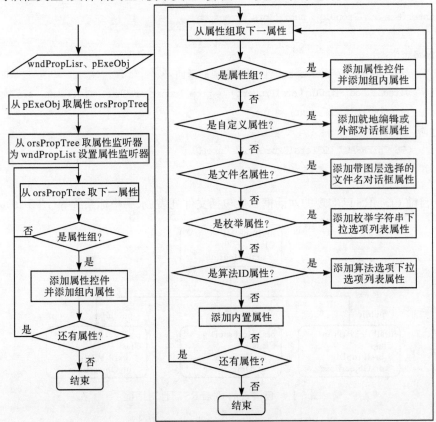

图 5-20 AddBCGPropertyTree 流程

5.3.3 OpenRS 属性事件的响应与动态属性界面

1. 动态属性界面需求

动态界面配置是指算法参数可以根据输入影像或者算法不同而自动地改变界面上的配置参数。属性的动态配置使界面更灵活，算法实现也得到简化。

如图 5-21 所示，以支持向量机的参数动态配置界面，说明动态配置的意义。根据所选择的核的类型(SvmKernelType)不同，将自动出现对应该核类型的配置参数。选择径向基函数核 RBF 时，配置参数为 gamma 系数。

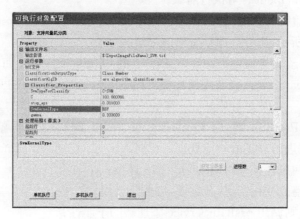

图 5-21　动态属性界面示例 1

如图 5-22 所示，选择多项式核 polynomial 时，配置参数除 gamma 系数外，还增加了系数 coef0 和多项式阶 degree。

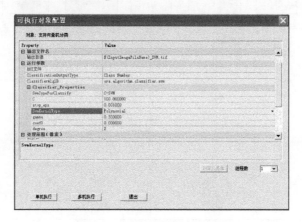

图 5-22　动态属性界面示例 2

2. OpenRS 的动态属性界面实现机制

OpenRS 的属性界面的动态性依赖于 CBcgPropList 的 OnPropertyChanged 回调函数：

```
virtual void OnPropertyChanged(CBCGPProp * pProp)const;
```

OpenRS 的界面属性列表控件 orsBcgPropList 重载了 CBcgPropList 的 OnPropertyChanged。基于回调函数 OnPropertyChanged 和委托技术，可以将属性修改事件传递到 OpenRS 可执行对象。

1) 事件的监听与处理流程

如图 5-23 所示，属性界面动态修改的流程如下。

（1）用户修改属性值，触发 CBCGPProp 的 OnUpdataValue 回调函数。

（2）OnUpdataValue 修改完属性值后回调 orsBcgPropList 的 OnPropertyChanged 函数。

（3）orsBcgPropList 的 OnPropertyChanged 回调可执行对象的 OnPropertyChanged 函数进行真正的响应处理：

① 首先取回属性控件对应的 OpenRS 属性；

② 分析属性的变化情况，若需要，则删除子属性；

③ 若需要，则调用 orsIGuiService 添加新的子属性；

④ 调用 orsBcgPropList 的 AdjustLayout 和 RedrawWindow 调整界面布局和窗口大小。

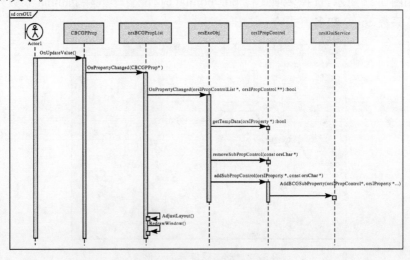

图 5-23　属性界面修改的顺序图

在 orsBcgPropList 中，属性改变回调函数把发生改变的属性控件传递给属性事件监听器 m_listener。具体代码如下：

```
void OnPropertyChanged(CBCGPProp *pProp)const
{
    if(NULL != m_listener)
    {
        //转换为属性控件接口
        orsIPropControl *pControl = reinterpret_cast<orsIPropControl *>(pProp
            ->GetData());

        orsIPropControlList *pPropControlList = ORS_PTR_CAST(orsIPropControl-
            List, this);

        if((*m_listener)(pPropControlList, pControl)){
            ((CBCGPPropList *)this)->AdjustLayout();
            ((CBCGPPropList *)this)->RedrawWindow();
        }
    }
}
```

这里的属性事件监听器 m_listener 实际上是一个委托 orsPropertyListener，对应于可执行对象的回调函数 OnPropertyChanged。由于使用了委托计数，这里的 m_listener 可以源于任何对象。

从 BCG 的属性获得 OpenRS 的属性控件指针时使用了 CBCGPProp 的一个数据指针，该指针指向继承 CBCGPProp 和 pPropControlList 的属性对象。该指针在调用 bindOrsProperty 时使用 SetData 设置。

2) 属性界面动态修改实例

下面以监督分类的属性界面回调来说明动态界面修改的具体过程。

(1) OpenRS 可执行对象中的属性修改函数定义。

在\OpenRS\desktop\include\orsBase\orsIProperty.h 中的委托定义：

```
using namespace fastdelegate;
typedef FastDelegate2<orsIPropControlList *, orsIPropControl *, bool> orsProperty-
    Listener;
```

在\OpenRS\desktop\src\plugins\orsImageClassify\orsXImageSourceClassify.h 中的委托变量定义：

```
class orsXImageSourceClassify: public orsIImageSourceHelper_prop<orsIImageSource-
    Classify>, public orsObjectBase
{
    ...
private:
    orsPropertyListener m_propListener;
}
```

在 orsXImageSourceClassify 构造函数中绑定委托函数 OnPropertyChanged，并作为属性添加到的属性树 sourceArguments 中。

```
orsXImageSourceClassify:: orsXImageSourceClassify ( bool bForRegister, orsChar
    *classifyAlgID):orsIImageSourceHelper_prop<orsIImageSourceClassify>(bForRegister)
{
    if(!bForRegister){
        ...
        //绑定成员函数
        m_propListener.bind(this, &orsXImageSourceClassify::OnPropertyChanged);
        //添加到属性 sourceArguments
        m_sourceArguments->addAttr(ORS_PROPERTY_LISTNER, &m_propListener);
    }
}
```

监督分类程序的编写采用了 orsIExecute、orsIImageSource 和 orsIAlgClassifier 三级的形式（参见 10.3 节）。其中，可执行对象的实现是通过模板 orsIExeImageSourceHelper、ORS_EXE_IMP2 和 orsXImageSourceClassify 实现的。orsXImageSourceClassify 以 m_classifyAlgID 为参数选择分类算法。图 5-24 给出了常用的分类算法对象。

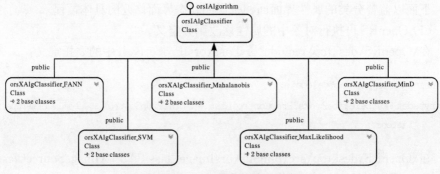

图 5-24　监督分类的算法对象

(2) 逐级回调处理。

发生属性修改时,orsBcgPropList 回调 orsXImageSourceClassify::OnPropertyChanged。orsXImageSourceClassify::OnPropertyChanged 检查分类算法是否发生了变化。若分类算法发生了变化,则通过调用 removeSubPropControl 删除旧的算法参数(属性),然后调用 addSubPropControl 添加新的算法参数(属性)。orsXImageSourceClassify 的 OnPropertyChanged 代码如下:

```cpp
#include "orsGuiBase/orsIPropControl.h"
bool orsXImageSourceClassify::OnPropertyChanged(orsIPropControlList *pCtrlList,
    orsIPropControl *pControlChanged)
{
    //控件状态初始化阶段,不需要响应
    if(NULL == pControlChanged)
        return false;

    ref_ptr<orsIProperty> tempData = getPlatform()->createProperty();
    //取变化控件的属性名
    orsString propName = pControlChanged->getPropName();
    //分类器 ID 属性?
    if(propName == classiferID){
        //取变化控件属性到临时变量
        pControlChanged->getTempData(tempData.get());

        orsString classifyAlgID;
        if(tempData->getAttr(classiferID, classifyAlgID)){//取分类算法 ID
            m_classifyAlg = ORS_CREATE_OBJECT(orsIAlgClassifier, classifyAlgID);
            if(m_classifyAlg.get()&& m_classifyAlg->getProperty()){
                ref_ptr<orsIProperty> pClassifierProp = m_classifyAlg->getProperty();
                //取分类器属性组
                //不能直接使用指针,否则会被当成 orsIObject
                m_parameterArgs->setAttr(classiferPROP, pClassifierProp);
                orsIPropControl *pParent = pControlChanged->getParent();
                //调整父属性的子属性
                if(pParent){
                    //删掉旧的分类器属性
                    pParent->removeSubPropControl(classiferPROP);
```

```cpp
            //添加新的分类器属性
            pParent->addSubPropControl(m_parameterArgs.get(), classi-
                ferPROP);
        }
    }
}

    return true;
}

//若不是分类算法发生变化,则调用分类算法的响应函数进行可能算法参数变化处理
{
    orsPropertyListener *pListener = NULL;
    m_classifyAlg->getProperty()->getAttr(ORS_PROPERTY_LISTNER, pListener);
    if(pListener)
        return( *pListener)(pCtrlList, pControlChanged);
}

    return false;
}
```

假设分类算法为 SVM(支持向量机),SVM 的核类型(kernelType)发生了变化,则 SVM 的其他参数要跟着发生变化。支持向量机的属性回调代码如下:

```cpp
bool orsXAlgClassifier_SVM::OnPropertyChanged(orsIPropControlList * pCtrlList,
    orsIPropControl *pControlChanged)
{
    //控件状态初始化,不需要响应?
    if(NULL == pControlChanged)
        return false;
    //取变换控件的属性名
    orsString propName = pControlChanged->getPropName();

    ref_ptr<orsIProperty> tempData = getPlatform()->createProperty();
    //核类型属性
    if(propName == kernalTYPE){
        //取变化控件属性到临时变量
```

```cpp
pControlChanged->getTempData(tempData.get());

orsString kernelType;
tempData->getAttr(propName, kernelType);//取核类型
//取父控件
orsIPropControl * pParent = pControlChanged->getParent();
//根据核类型,增删该类核所需参数
if(kernelType == kernalTYPE_L)
{
    pParent->removeSubPropControl("gamma");
    pParent->removeSubPropControl("coef0");
    pParent->removeSubPropControl("degree");
}
else if(kernelType == kernalTYPE_R)
{
    pParent->removeSubPropControl("degree");
    pParent->removeSubPropControl("coef0");
    m_parameterArgs->addAttr("gamma", m_svmPar.gamma, true);
    pParent->addSubPropControl(m_parameterArgs.get(), "gamma");
}
else if(kernelType == kernalTYPE_S)
{
    pParent->removeSubPropControl("degree");
    m_parameterArgs->addAttr("gamma", m_svmPar.gamma, true);
    m_parameterArgs->addAttr("coef0", m_svmPar.coef0, true);
    pParent->addSubPropControl(m_parameterArgs.get(), "gamma");
    pParent->addSubPropControl(m_parameterArgs.get(), "coef0");
}
else if(kernelType == kernalTYPE_P)
{
    m_parameterArgs->addAttr("gamma", m_svmPar.gamma, true);
    m_parameterArgs->addAttr("coef0", m_svmPar.coef0, true);
    m_parameterArgs->addAttr("degree",(ors_int32)m_svmPar.degree, true);

    pParent->addSubPropControl(m_parameterArgs.get(), "gamma");
    pParent->addSubPropControl(m_parameterArgs.get(), "coef0");
```

```
                pParent->addSubPropControl(m_parameterArgs.get(), "degree");
            }
        }

        return true;
    }
```

完整代码见\OpenRS\desktop\src\plugins\orsImageClassify 目录下的 orsXImageSourceClassify.h、orsXImageSourceClassify.cpp、orsXAlgClasifier_SVM.h 和 orsXAlgClasifier_SVM.cpp。

5.4 语言本地化

5.4.1 MFC 的本地化方法

在软件国际化的今天,资源从代码中独立出来,使在不同语言操作系统下能运行不同语言版本的程序,是很有意义的事(昌燕,2006)。

在 VC6.0 及以前的版本中,需要在 CWinApp::InitInstance 加入载入语句:

```
LoadLibrary("ResourceDll.dll");
```

但是在 VC6.0 以后的 MFC 版本中,已经加入了自动载入资源 DLL 的功能。只需要将 DLL 文件命名为"程序名+[语言名].dll"即可。MFC 自动检测当前系统所使用的语言环境,然后尝试载入相应的资源 DLL。若 app.exe 内置资源是英文的,且同一目录下有 appCHS.dll 和 appCHT.dll,则运行 app.exe 时,会根据系统的语言载入相应的 DLL。若系统是繁体的,则自动载入 appCHT.dll;若是简体的,则载入 appCHS.dll;若都不符合,则使用内置的资源。

MFC 7.0 及更高版本提供对附属 DLL 的增强支持,该功能有助于创建针对多种语言进行本地化的应用程序。附属 DLL 是一个纯资源 DLL,它包含应用程序针对特定语言进行本地化的资源。当应用程序开始执行时,MFC 自动加载最适合于环境的本地化资源。

使用 VS2005 可以很方便地创建包含资源的 DLL。下面简要介绍其步骤。

(1) 新建一个与代码文件夹平级的文件夹,文件夹的名字为:代码程序的工程名字+Res。

(2) 将所有的资源文件(*.rc)、位图、图标文件和 resource.h 复制到资源文件夹下。

(3) 打开 VS2005 选择 File\New\Project From Existing code,选择工程的种

类为 VC++,将工程文件的位置选定为资源文件夹,工程名称为资源文件夹的名称,Next 后选择工程类型为 DLL 工程,然后一路 Next 下去直到完成。

5.4.2 OpenRS 的本地化方案

考虑到 OpenRS 要支持 VC6 开始的所有 Visual C++ 版本,同时可以动态地实现本地化的修改,OpenRS 设计了一套简单易用的本地化方案。该方案以英文版本作为内置语言,运行时通过本地化文件进行文本替换实现本地化,支持中文和其他语言。

OpenRS 本地化作为平台机制,可以自动实现插件菜单的本地化。其本地化的机制是通过一个"程序内部文本"与"本地文本"的对应文件实现的。该文件放在目录 etc\localization 下,如中文的本地文件为 etc\localization\cn\localString.txt。查找文本中以":="作为分隔符的字符串对就可以实现文本的替换。并且由于没有采用空格相关的分隔符,因此文本字符串可以带空格。示例如下:

```
Vector := 矢量
Workflow := 工作流
LandProduction := 陆地产品
orthoTrans := 正交变换
imageRegister := 影像配准
imageSegmentation := 影像分割
imageFilter := 影像滤波
imageClassify := 影像分类
imageHyper := 高光谱
Digitize Point := 数字化点
```

5.4.3 OpenRS 的本地化的实现

```
const orsChar * GetLocalizedString(const orsChar * englishStr);
```

输入一个字符串,返回本地化字符串。若没有对应的本地化字符串,则返回原字符串。具体实现参见\OpenRS\desktop\src\orsGuiBase\orsXGuiLocalizationService.cpp。

第 6 章　桌面集成环境设计与实现

6.1　OpenRS 主控模块

OpenRS 主控模块的主要功能是实现 OpenRS 相关软件模块的快速集成。OpenRS 主控模块在功能上模仿 ERDAS，在实现上借鉴 Windows 桌面的实现机制。如图 6-1 所示，OpenRS 主控模块的图标其实就是执行程序快捷方式。快捷方式放在 OpenRS 执行程序目录下的 iconPanel 子目录中，即可出现在面板上。

图 6-1　OpenRS 主控模块界面

除了集成执行程序，OpenRS 主控模块还提供了对象查询、插件查询、RDF 查询功能。如图 6-2 所示，对象查询按对象类别树列出所有的对象。对于每个对象，可以查询对象所在的插件。如果不知道一个对象的 ID，则可以在对象树上查找到该对象，然后复制出该对象的 ID。

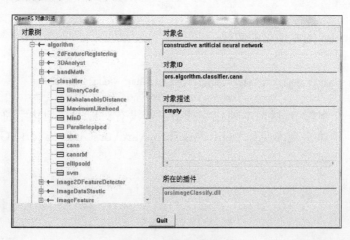

图 6-2　对象查询界面

如图 6-3 所示，插件查询列出所有已注册的插件。对于每个插件，可以查询插件所包含的对象。

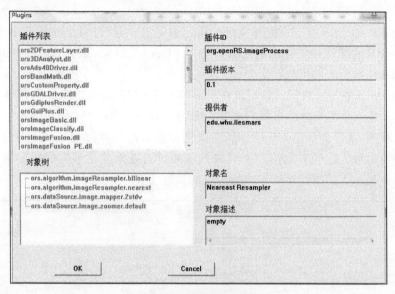

图 6-3　插件查询界面

如图 6-4 所示，RDF 查询可以列出所有的资源描述三元组。

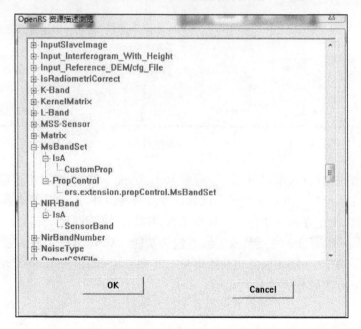

图 6-4　RDF 查询界面

6.2 对象执行器——orsExeRunner

orsExeRunner 用于查询和执行可执行的对象（可执行对象参看 7.1 节）。如果 OpenRS 的插件对象查询能查询所有的 OpenRS 对象，那么 orsExeRunner 可以查询和执行所有的可执行对象。

OpenRS 中的可执行对象可以在不同的程序中调用，其中对象执行器 orsExeRunner 是一个用于测试可执行对象的程序。orsExeRunner 是一个基于对话框的执行程序，主界面如图 6-5 所示。左侧窗口显示的是当前 OpenRS 平台中可用的可执行对象。这些对象按照类别关系即接口的继承关系排列成树形结构。右侧窗口中显示当前所选定的可执行对象的参数配置情况。

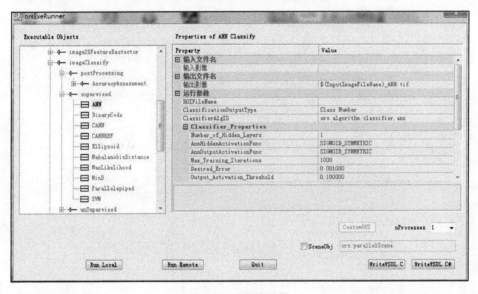

图 6-5 对象执行器

在设置好参数后，单击 Run Local 或 Run Remote 按钮，即可将选定的可执行对象插件在本地执行或远程执行。除算法本身的参数外，还可以指定运行时 CPU 进程数，可以指定在本地执行还是远程执行，并指定并行执行场景对象。

另外，orsExeRunner 还集成了服务包装功能。WriteWSDL 按钮提供了一键式服务包装功能（一键式服务包装原理见第 9 章），生成用于 IIS 的 C#代码或用于 OpenRS 专用服务器的 C 代码（图 6-5）。

6.3 基于图层的影像、矢量显示

6.3.1 基本显示架构

如图 6-6 所示，OpenRS 采用图层化显示架构，实现影像、矢量及任意扩展图层的显示。图层化的显示界面能够很好地集成栅格、矢量等不同内容的显示，可以方便地实现图层间的卷帘、假彩色组合等方面实用的图像比较工具。OpenRS 图层化显示架构以 ArcMap 的图层显示架构为基础进行自主开发，在图层树的每个图层显示图标上嵌入栅格、矢量图层的定制功能，避免采用统一的灰度拉伸工具条等容易混淆的方式。

图 6-6 OpenRS 图层化显示界面

1. 图层化显示架构

在显示视图刷新或暴露时，视图绘制函数根据显示空间在视图中的当前位置，对后备位图进行调整，若发现有需要绘制的位图，则调用图层树管理器的绘制函数对后备位图进行绘制。图层树管理器的绘制函数则逐层检查图层树上的图层，逐一在后备位图进行绘制。这一过程可以考虑图层之间的遮挡、透明等情况进行综合优化。

图 6-7 为 OpenRS 图层化显示架构。其中，显示视图提供显示的窗口，后备位图作为显示缓存，图层管理器是图层的容器，图层树界面按图的方式对图层进行可视化管理，图层渲染器在缓存上进行绘制。

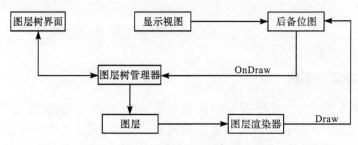

图 6-7　OpenRS 图层化显示基本架构

在显示视图刷新或暴露时，视图绘制函数根据显示空间在视图中的当前位置，对后备位图进行调整，若发现有需要绘制的位图，则调用图层树管理器的绘制函数对后备位图进行绘制。而图层树管理器的绘制函数则逐层检查图层树上的图层，逐一在后备位图进行绘制。这一过程可以考虑图层之间的遮挡、透明等情况进行综合优化。

2. 视图的缓存机制

OpenRS 的快速演示模块 Fastdisplay 实现了基于缓存的快速显示架构（图 6-8）。在 Fastdisplay 的 BackbufferDC 中把显示空间和缓存都划分为 256×256 小块。缓存分配的原则比显示窗口的大两圈，即保证显示窗口在影像上漫游时能够保证有一圈的富余。

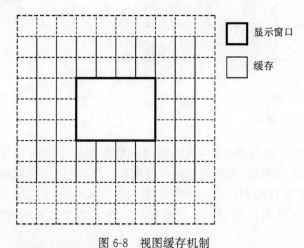

图 6-8　视图缓存机制

缓存更新机制如下：
(1) 根据当前窗口计算需要装载的影像分块范围。
(2) 把界外的缓存块编号压入空闲块堆栈。

(3) 直接读取显示急需的影像块。

(4) 更新预读影像列表。

3. 单个影像的渲染与显示

OpenRS 的显示是专门为影像显示优化的，因此属于以影像为主的显示，矢量的显示最后都要转换到影像坐标进行显示。影像的显示采用影像处理链的方式，通过影像源对象的组合，实现影像的拉伸、放大等操作。

如图 6-9 所示，对于单独一个影像的显示，一般需要经过亮度拉伸和几何放大两个环节。

图 6-9　基本影像显示

6.3.2　多视图显示的需求分析

OpenRS 的图层化显示设计目标是能够适用于多个视图的情况，而不只局限于一个视图。例如，一般情况下影像的配准需要两个视图，而空三编辑则需要三个以上视图。

表 6-1 给出了不同应用情况下对显示的可能需求，涉及两个方面的内容：第一个方面和窗口有关，有些图层的内容只能在某一个特点的窗口显示，而有些图层的内容必须在各个窗口中同时显示；第二个问题和坐标变换有关。对于栅格、矢量的多图层显示，由于各个图层坐标参考系可能不一样，因此必须在显示时进行转换。

表 6-1　不同应用情况下对显示的可能需求

用例	图层内容	窗口需求	二维矢量	三维矢量
影像处理	原始影像层、处理结果影像层	一个二维窗口	无	无
地物提取	原始影像层、提取矢量层	一个二维窗口	ROI、影像分割结果等	外部导入的矢量等
立体测图	左影像层、右影像层、测图结果层	一个立体窗口	半自动测图中间结果匹配的点线面匹配的视差格网	结果矢量
空三选点	空三所有影像层、连接点图层	多个二维窗口 一个立体窗口	连接点	地面控制点

按照惯例，一般都会选择每个窗口中第一个出现的图层作为参考，其他图层的显示需要和第一个图层自动对齐。为了简单起见，OpenRS 中规定每一个显示窗口的第一个显示图层必须是栅格影像，称为视图空间（ViewSpace）。所有后续

打开图层的显示都需要先转换到 ViewSpace 给定的参考坐标系。

1. 一个视图中多个影像图层的渲染与显示

对于多个影像的显示,除第一个影像外,其他影像必须经进行动态几何变换,先把需要显示的影像和第一个影像进行空间对准,才能进行正常显示。如图 6-10 所示,OpenRS 采用一个影像 Warper 对影像进行动态空间变换,来对准不同空间参考或几何模型的影像。因为影像 Warper 中已经包含缩放功能,所以影像放大器可以省略。

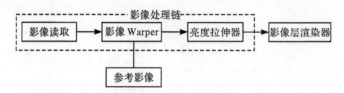

图 6-10　基于影像 Warper 的多影像层叠加显示原理

2. 一个视图中矢量图层的渲染与显示

OpenRS 的图层显示是以影像为基础的。矢量的显示先要把坐标转换到参考影像作为参照。转换的环节可能会涉及不同空间参考坐标的转换,三维坐标到二维坐标的转换等。目前是根据一个矢量图层是否具有空间参考来确定的。如果一个矢量图层具有空间参考,那么认为它的矢量坐标是三维影像坐标,否则认为是二维地理坐标。

如果是二维地理坐标,那么可以直接进行缩放显示。

如果是三维地理坐标,那么首先需要把坐标转换到 ViewSpace 的空间参考系统,然后进行缩放。

3. 不同图层在多个视图中的渲染与显示

对于需要在不同视图中显示的图层,OpenRS 采用视图绑定的方式来说明该图层是否需要在某个视图中显示。简单地说,就是每个图层对象内置了视图指针数组用于保存绑定的视图指针。

6.3.3　多图层、多视图快速显示架构

1. 图层绘制的参考视图空间

如图 6-11 所示,图层绘制的视图空间定义了物方三维坐标的空间参考、物方三维坐标到二维影像坐标的成像几何模型,以及视图本身的平移、缩放等参数。任何

与视图空间具有明确二维、三维坐标关系的图层都可以在视图上进行绘制。这里的视图是一个广义的概念，可以是实际的绘制窗口，也可以是内存位图等虚拟设备。

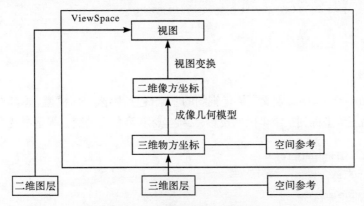

图 6-11　图层绘制的视图空间

orsImage2View 定义了二维影像到视图的缩放和平移关系。image2View 和 view2Image 给出了影像坐标和视图坐标的转换关系。

```
//影像坐标到视图坐标的变换
struct orsImage2View{
public:
    float m_zoom;
    double m_offsetX, m_offsetY;

public:
    orsImage2View():
        m_zoom(0.0), m_offsetX(0.0), m_offsetY(0.0){};

    orsImage2View(float zoomX, float zoomY, double offsetX, double offsetY):
        m_zoom(zoomX), m_offsetX(offsetX), m_offsetY(offsetY){};

    void image2View(orsPOINT2D *pt)const
    {
        pt->x = m_offsetX + m_zoom * pt->x + 0.5;
        pt->y = m_offsetY + m_zoom * pt->y + 0.5;
    }

    void view2Image(orsPOINT2D *pt)const
```

```
        {
            pt->x = (pt->x - m_offsetX)/m_zoom;
            pt->y = (pt->y - m_offsetY)/m_zoom;
        }
        ...
};
```

orsIImageViewBase 定义了影像视图的基础,包括影像几何模型、基本的影像几何关系,包括三维坐标到二维坐标的投影、二维坐标和高程面的交会等基本变换功能。

```
struct orsImgViewInfo{
    double xl, yt;
    double xr, yb;
    double zoom;
};

interface orsIImageViewBase
{
public:
    virtual bool GetImgViewInfo(int viewIndex, orsImgViewInfo * viewInfo) = 0;
    virtual void SetViewSpace(orsIImageSource * pImg) = 0;
    virtual const orsIImageSource * GetViewSpace() = 0;
    virtual orsIImageGeometry * GetImageGeometry() = 0;
    virtualvoid SetUserSRS(orsISpatialReference * pSRS) = 0;

    virtual void Project(const orsPOINT3D &xyz, orsPOINT2D * xy) = 0;
    virtual void IntersetWithZ(const orsPOINT2D * pts 二维, double z , int numofPts,
        orsPOINT3D * pts3d) = 0;

    virtual void SetLayerCollection(orsILayerCollection * layerCollection) = 0;
    ...
};
```

orsIImageView 和绘制窗口对应,包含了主要的二维绘制函数和三维绘制函数。orsIImageViewBase 从 orsIImageView 中独立出来的原因照顾到在立体显示一个视图中包含了左右两个影像绘制空间的情况。而 orsIImageView 的二维绘制函数都要求指定是哪一个视图空间。

一般来说,orsIImageView 除了传递给图层进行绘制,更重要的是传递给需要进行交互式操作的界面扩展对象。

对于具有交互式操作的界面扩展对象,可以通过 SetActiveViewEventListener 设置监听器,捕获鼠标、键盘等消息。

```
interface orsIImageView: public orsIImageViewBase
{
public:
    //使区域无效, pRect = NULL 为全部
    virtual void InvalidateRect(orsInvalidateTYPE eInvalidType, const orsRect_i *
        pRect = NULL) = 0;


    //在 zoomIn, zoomOut、pan 等和鼠标相关的视图监听器激活时,设置相应的监听器
    virtual void SetActiveViewEventListener(orsIViewEventListener * listner) = 0;
    //取当前激活的监听者,可用于设置菜单、工具条按钮的 check 状态
    virtual const orsIViewEventListener * GetActiveListner()const = 0;

public:
    virtual void HideCursor() = 0;
    virtual void ShowCursor() = 0;

    //滚动到指定的影像位置
    virtual void DriveTo(const orsPOINT2D &pts2D) = 0;
    virtual void DriveTo(const orsPOINT3D &pts3d) = 0;

public:
    virtual bool PrepareHDC(int nPenStyle, float fWidth, COLORREF crLColor, bool
        bFill, int nHatchIndex, COLORREF crFillColor) = 0;

    virtual bool EnableXorMode(bool bEnbale = true) = 0;
    virtual bool EnableTransparent(bool bTransparent = true) = 0;

public://二维绘制方法,坐标为影像坐标
    virtual void DrawPoint(int viewIndex, const orsPOINT2D &pPoint, float fAngle,
        int r) = 0;
    virtual void DrawPolyline(int viewIndex, const orsPOINT2D * pPoint, long nCount,
        bool bClosed = false) = 0;
    virtual void DrawPolygon(int viewIndex, const orsPOINT2D * lpPoints, int nCount)
        = 0;
    virtual void DrawPolyPolygon(int viewIndex, const orsPOINT2D * pPoint, const
```

```cpp
        long * lpPolyCounts, long nCount) = 0;
    virtual void DrawArc(int viewIndex, const orsPOINT2D &topLeft, const orsPOINT2D
        &bottomRight, const orsPOINT2D &ptStart, const orsPOINT2D& ptEnd) = 0;
    virtual void DrawArc(int viewIndex, const orsPOINT2D &p1, const orsPOINT2D &p2,
        const orsPOINT2D &p3) = 0;

    virtual void DrawCircle(int viewIndex, const orsPOINT2D& ptTopLeft,   const or-
        sPOINT2D& ptBottomRight) = 0;
    virtual void DrawRectangle(int viewIndex, const orsPOINT2D& ptTopLeft,   const
        orsPOINT2D& ptBottomRight) = 0;
    virtual void DrawRoundRect(int viewIndex, const orsPOINT2D&  ptTopLeft,   const
        orsPOINT2D& ptBottomRight, const orsPOINT2D& ptRadius) = 0;
    virtual void DrawEllipse(int viewIndex, const orsPOINT2D& ptTopLeft,   const or-
        sPOINT2D& ptBottomRight, float fAngle) = 0;

    virtual void DrawText(int viewIndex, const orsPOINT2D &Point, const orsChar *
        lpszString, int nCount, const LOGFONT * pLogFont, COLORREF crForeColor, COL-
        ORREF crBackColor) = 0;
    virtual void DrawText2(int viewIndex, const orsPOINT2D * pPoint, const float *
        fAngle, const orsChar * lpszString, int nCount, const LOGFONT * pLogFont,
        COLORREF crForeColor, COLORREF crBackColor) = 0;

public: //三维绘制方法,坐标为视图空间坐标
    virtual void DrawPoint(const orsPOINT3D &pPoint, float fAngle, int r) = 0;
    virtual void DrawPolyline(const orsPOINT3D * pPoint, long lPointCount, bool
        bClosed = false) = 0;
    virtual void DrawPolygon(const orsPOINT3D * lpPoints, int nCount) = 0;
    virtual void DrawPolyPolygon(const orsPOINT3D * pPoint, const long * lpPolyCou-
        nts, long nCount)   = 0;

    virtual void DrawArc(const orsPOINT3D &topLeft, const orsPOINT3D &bottomRight,
        const orsPOINT3D &ptStart, const orsPOINT3D& ptEnd) = 0;
    virtual void DrawArc (const orsPOINT3D &p1, const orsPOINT3D &p2, const or-
        sPOINT3D &p3) = 0;
    virtual void DrawCircle(const orsPOINT3D& ptTopLeft,   const orsPOINT3D& ptBot-
        tomRight) = 0;
    virtual void DrawRectangle(const orsPOINT3D& ptTopLeft, const orsPOINT3D& pt-
        BottomRight) = 0;
    virtual void DrawRoundRect(const orsPOINT3D&   ptTopLeft,    const orsPOINT3D&
```

```
    ptBottomRight, const orsPOINT3D& ptRadius) = 0;
virtual void DrawEllipse(const orsPOINT3D& ptTopLeft,  const orsPOINT3D& ptBot-
    tomRight, float fAngle) = 0;

virtual void DrawText(const orsPOINT3D &Point, const orsChar * lpszString, int
    nCount, const LOGFONT * pLogFont, COLORREF crForeColor, COLORREF crBackCol-
    or) = 0;
virtual void DrawText2(const orsPOINT3D * pPoint, const float * fAngle, const
    orsChar * lpszString,  int nCount, const LOGFONT * pLogFont, COLORREF
    crForeColor, COLORREF crBackColor) = 0;

virtual void SetImage(orsIImageSource * imgCache) = 0;
};
```

2. 图层与图层树

与 ArcMap 一样，OpenRS 图层树的管理包括图层组、图层两种基本的图层树元素，而 layerSource 是一种特殊的图层组，专门用于表示来自同一个目录或文件的多个图层。

图 6-12(a)和(b)给出了在图层树视图和数据源视图下的图层树显示方式。图层树视图按照图层分组关系，以树的形式给出了两个影像图层和一个矢量图层组，矢量图层组下又包含了三个矢量图层。数据源视图按照图层所在的目录进行分层显示。

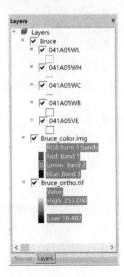

(a) 图层树　　　　　　　(b) 数据源树

图 6-12　图层树与数据源

1) 图层树元素接口

图层树元素除了作为图层组和图层的父类,很重要的是要为图层树控件的节点和图层树节点提供关联。从这个意义上说,图层树控件为图层树提供了一个可视化视图,图层树元素接口定义了图层树元素在图层树控件上的行为(图 6-13)。

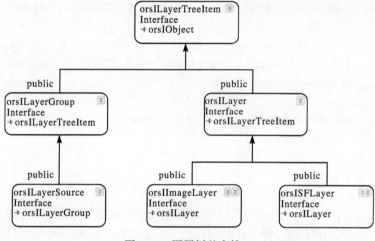

图 6-13　图层树基本接口

为了关联图层树控件的节点,定义了 orsTreeItemHandle。orsTreeItemHandle 其实就是为了避开图层树界面实现的一个句柄。

```
typedef void * orsTreeItemHandle;
```

在采用 MFC 界面框架的情况下,orsTreeItemHandle 就是 MFC 的 HTREEITEM。

orsLayerStyleITEM 用于保存图层风格。

```
struct orsLayerStyleITEM{
    char tagStr[256];
    orsTreeItemHandle hTreeItem;
};
```

orsLayerTreeItemTYPE 枚举了四种图层树元素,用于进行图层树节点的渲染。

```
enum orsLayerTreeItemTYPE{
    ORS_LTT_LAYER = 0,
    ORS_LTT_LAYER_GROUP = 1,
```

```
    ORS_LTT_LAYER_SOURCE = 2,
    ORS_LTT_LAYER_SOURCE_DIR = 2
};
```

orsILayerTreeItem 定义了图层元素的树控件绑定、可见性设置、预绘制调用等功能。预绘制 OnPreDraw 根据当前的绘制参数,返回需要绘制的图层列表。

```
interface orsILayerTreeItem : public orsIObject
{
public:
    //提供给树控件使用,用于关联图层节点和树控件节点
    orsTreeItemHandle m_hLayerTreeItem;
    orsTreeItemHandle m_hLayerSourceTreeItem;

    orsILayerTreeViewer * m_pLayerTreeViewer ;
    orsILayerTreeViewer * m_pDataSourceTreeViewer ;

public:
    orsILayerTreeItem()
    {
        m_hLayerTreeItem = NULL;
        m_hLayerSourceTreeItem = NULL;
        m_pLayerTreeViewer = NULL;
        m_pDataSourceTreeViewer = NULL;
    };

    //取图层类型
    virtual int layerTreeItemType() = 0;

    //取树控件上的树节点的绘制风格项
    virtual orsLayerStyleITEM * GetStyleItemsOnTreeCtrl(int &nItems) = 0;

public:
    virtual bool IsVisible() = 0;
    virtual void SetVisible(bool bVisible, bool bSetChildren = false) = 0;

    virtual orsILayerGroup * GetParent() = 0;
    virtual void SetParent(orsILayerGroup * parent) = 0;
```

```cpp
    virtual const orsChar * GetLayerName() = 0;
    virtual void SetLayerName(const orsChar * layerName) = 0;

public:
    //从上往下预绘制,计算每个图层的裁剪区,并从总裁剪区去除
    virtual void OnPreDraw(orsDC_TYPE eDcType, orsIImageView * view, orsImage2View
        * map2View, orsIRegion * clipRgn, orsArray〈orsILayer * 〉 &layersToBeDrawn)
        = 0;
    //OpenGL 三维视图绘制
    virtual void OnDraw3D() = 0;

    //选取图层对象,如矢量、像素等
    virtual bool PickObject(int viewIndex, orsPOINT2D &imgPt, orsPOINT3D &point3D)
        = 0;

    ORS_INTERFACE_DEF(orsIObject, "layerTreeItem")
};
```

2) 图层接口

图层接口在图层树元素的基础上定义了图层的绘制行为,包括空间参考的获取、图层与视图的绑定、图层内容的绘制 OnDraw、树控件上图层风格的绘制 OnDrawStyle 等。

```cpp
interface orsILayer : public orsILayerTreeItem
{
public:
    virtual void NotifyUpdataLayerRender() = 0;
public:
    virtual bool IsValid()=0;
    virtual const orsChar * GetLayerType()=0;

    virtual orsRect_i GetAreaOfInterest()=0;
    virtual orsRect_i GetExtent()=0;

    virtual orsISpatialReference * GetSpatialReference() = 0;

    //若 imageView 不为空,则该层为影像层,坐标为影像坐标
    virtual orsIImageView * GetImageViewBoundTo() = 0;
```

```cpp
//把该图层绑定到具体的视图(该层为影像层,坐标为影像坐标)
virtual bool BindImageView(orsIImageView * pImageView) = 0;

virtual void SetLayerSource(orsILayerGroup * pLayerSource) = 0;
virtual orsILayerGroup * GetLayerSource() = 0;

public:
    //设置对话框
    virtual void SetRender(int flag) = 0;

    //在图层上绘制,首先需要调用 PreDraw 计算裁剪区
    virtual void OnDraw(orsDC_TYPE eDcType, orsHDC dc, orsIImageView * view,
        orsImage2View * map2View) = 0;

    //在 TREE 上绘制 STYLE
    virtual void OnDrawStyle(orsHDC dc, orsRect_i &rect, int flag) = 0;

public:
    ORS_INTERFACE_DEF(orsIObject, "layer");
};
```

3) 图层组接口

图层组对象负责管理一组图层,主要为图层的添加、删除、插入、绘制传递等。

```cpp
interface orsILayerGroup : public orsILayerTreeItem
{
public:
    virtual void GetLayerNames (const orsChar * layerTypeName, orsArray<const
        orsChar *> &layerFileNames) = 0;
    virtual orsILayerTreeItem * GetLayerByName(const orsChar * layerName) = 0;

    //在 TREE 上绘制 STYLE
    virtual void OnDrawStyle(orsHDC dc, orsRect_i &rect){}
    //绘制选中的像素、矢量等
    virtual void DrawSelectedObject(orsHDC dc, orsRect_i &rect){}

    //在尾部添加
    virtual void AddItem(orsILayerTreeItem * item) = 0;
    //插入
```

```
virtual void InsertItem(orsILayerTreeItem * item, orsILayerTreeItem * pInser-
    tAfter) = 0;
virtual void RemoveItem(int index) = 0;
virtual void RemoveItem(orsILayerTreeItem * item) = 0;

//drived from orsILayerTreeItem
virtual int GetNumOfChildren() = 0;
virtual orsILayerTreeItem * GetChildren(int index)=0 ;
virtual int GetChildIndex(orsILayerTreeItem * pChildItem) = 0;
};
```

图层很重要的一个功能是组内图层的绘制,该功能通过通知递归调用组内图层或子图层组,收集需要绘制的图层,并加入 layersToBeDrawn。

```
void orsXLayerGroup::OnPreDraw(orsDC_TYPE eDcType, orsIImageViewBase * view, orsIm-
    age2View * map2View, orsIRegion * clipRgn, orsArray<orsILayer *> &layersToBeDrawn)
{
    int i;

    //图层组不可见?
    if(!m_bVisible)
        return ;

    for(i=0; i<m_children.size(); i++)
    {
        m_children[i]->OnPreDraw(eDcType, view, map2View, clipRgn, layersToBe-
            Drawn);
    }
}
```

3. 图层管理器和图层树视图

图层管理器是一个树状图层容器,负责图层的添加、删除、当前图层的设置、图层的绘制等;图层树视图接口定义了图层树控件的基本行为。当应用程序调用图层管理器添加删除图层时,图层管理器通过图层树视图接口函数通知图层管理器更新视图控件。反之,图层树控件上的图层位置发生变化时,图层树控件通过图层管理器接口对图层位置进行调整(图6-14)。

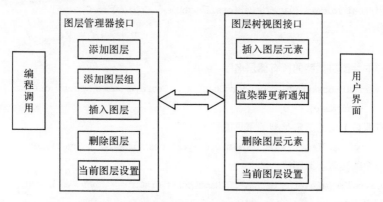

图 6-14　图层管理器和图层树视图的接口关系

1) 图层管理器

interface orsILayerCollection: public orsIObject
{
public:
　　virtual ～orsILayerCollection(){};

　　virtual void OnDraw(orsDC_TYPE eDcType, orsHDC hDC, orsIImageView * view, orsImage2View * map2View, orsIRegion * clipRgn) = 0;

public:
　　//virtual void SetLayerTree(orsILayerGroup * pLayerGroup) = 0;
　　virtual orsILayerGroup * GetLayerTree() = 0;
　　virtual orsILayerGroup * GetDataSourceTree() = 0;

public:
　　//设置图层树视图
　　virtual void SetLayerTreeViewer(orsILayerTreeViewer * pLayerTreeViewer)=0;
　　virtual orsILayerTreeViewer * GetLayerTreeViewer()=0;

　　//设置数据源树视图
　　virtual void SetDataSourceTreeViewer (orsILayerTreeViewer * pDatasourceTreeViewer)=0;
　　virtual orsILayerTreeViewer * GetDatasourceTreeViewer()=0;

　　//增加绘制视图,用于刷新视图

```cpp
    virtual void AddImageViewer(orsIImageView * pView) = 0;

public:
    //取某图层类型的所有图层名字
    virtual orsArray〈const orsChar *〉GetLayerNames (const orsChar * layer-
        TypeName) = 0;
    //取最上层的给定类型的可见图层
    virtual orsILayer * GetTopVisibleLayer(const orsChar * layerTypeName) = 0;
    //按图层名字取图层
    virtual orsILayer * GetLayerByName(const orsChar * layerName) = 0;
    //取当前 Simple Feature 图层
    virtual orsISFLayer * GetCurrentSFLayer() = 0;
    //取当前 Simple Feature 数据源层
    virtual orsISFLayerSource * GetCurrentSFLayerSource() = 0;

public:
    //往数据源树上加一个数据源
    virtual void AddDataSource(orsIDataSource * pDataSource, orsILayerTreeItem *
        pLayerTreeItem) = 0;
    //增加图层组
    virtual void AddLayerGroup(orsILayerGroup * pLayerGroup, orsILayerGroup * pPar-
        entGroup = NULL) = 0;
    //从数据源树上删除一个数据源
    virtual void RemoveDataSource(orsILayerTreeItem * pLayerTreeItem, bool bFrom-
        Layer = true) = 0;
    //从图层树中删除给定图层(不是物理删除)
    virtual void RemoveLayer(orsILayerTreeItem * pdelitem, bool bFromLayerSource =
        true) = 0;

    //在图层树顶部增加一图层
    virtual void AddLayer(orsILayer * pLayer, orsILayerGroup * pParentGroup =
        NULL) = 0;


    //在图层组(pLayerGroup)的图层项之前(pLayerBefore)插入图层
    //pInsertAfter = NULL 或 ORS_TVI_FIRST,插入最前
    //pInsertAfter = ORS_TVI_LAST, 插入最后
    virtual void InsertLayer(orsILayerGroup * pLayerGroup, orsILayerTreeItem * pIn-
        sertLayer, orsILayerTreeItem * pInsertAfter = (orsILayerTreeItem * )ORS_
```

```
        TVI_LAST) = 0;
    //把图层移动到图层组(pLayerGroup)的图层项之前(pLayerBefore), pInsertAfter 为 NULL
    //则在最前
    virtual void MoveLayer(orsILayerTreeItem * pSrcLayer, orsILayerGroup * pDst-
        Group, orsILayerTreeItem * pInsertAfter = (orsILayerTreeItem * )ORS_TVI_
        LAST) = 0;
    //删除所有图层
    virtual void RemoveAllLayers() = 0;

public:
    //按照图层名设置当前图层,并在视图上高亮
    virtual void SetCurrentLayer(const orsChar * layerName) = 0;
    //直接设置当前图层
    virtual void SetCurrentLayer(orsILayerTreeItem * pItem) = 0;
    //设定当前数据源
    virtual void SetCurrentLayerSource(orsILayerTreeItem * pItem) = 0;
    virtual orsILayerTreeItem * GetCurrentLayer() = 0;
    virtual orsILayerTreeItem * GetCurrentLayerSource() = 0;
public:
    ORS_INTERFACE_DEF(orsIObject, "layerCollection");
};
```

2) 图层树视图接口

```
struct orsILayerTreeViewer
{
public:
    virtual bool InsertDataFrame(orsIDataFrame * pDataFrame) = 0;

    virtual bool InsertLayerTreeItem(orsILayerGroup * pParentLayerGroup,
        orsILayerTreeItem * pInsertLayer, orsILayerTreeItem * pInsertAfter) = 0;
    virtual void DeleteLayerTreeItem(orsILayerTreeItem * pLayer) = 0;
    //通知树,更新 LayerTreeItem 的 Render 节点
    virtual void NotifyUpdataRenderItem(orsILayerTreeItem * pLayer) = 0;
    virtual void SetCurrentLayer(orsILayerTreeItem * pLayer) = 0;


};
```

6.3.4 影像图层及渲染

实现全色影像、多波段影像的显示,涉及的对象包括影像图层对象,影像层渲染器对象。

1. 影像层

影像层用于管理要显示的影像,并通过影像渲染器 orsIImageRender 提供影像显示功能。orsXImageLayer 的主要功能包括影像层的数据源设置、影像绘制、影像显示风格绘制等。

```
class orsXImageLayer : public orsILayerHelper<orsIImageLayer>, public orsObjectBase
{
…
public:
    orsXImageLayer(bool bForRegister);
    virtual ~orsXImageLayer();

    //返回图层类型
    virtual const orsChar * GetLayerType(){return ORS_LAYER_IMAGE; };

    //设置要显示的影像数据源
    virtual void SetImageSource(orsIImageSource * pImageSource, bool bAutoZoom =
        false);
    //直接打开要显示的影像
    virtual bool OpenImage(const char * imageFileName, orsILayerCollection * pLay-
        erCollection, orsIImageSource * viewSpace);

public:
    //预绘制,若需要绘制本影像,则把本图层添加到 layersToBeDrawn
    virtual void OnPreDraw(orsDC_TYPE eDcType, orsIImageViewBase * view, orsIm-
        age2View * map2View, orsIRegion * clipRgn, orsArray<orsILayer *>
        &layersToBeDrawn);
    //绘制本影像
    virtual void OnDraw(orsDC_TYPE eDcType, orsHDC hDC, orsIImageViewBase * view,
        orsImage2View * image2View);
    //绘制显示风格
    virtual void OnDrawStyle(orsHDC dc, orsRect_i &rect, int flag, orsRenderMode
```

```
        renderMode=ORS_RM_ONTREE);
    ...

public:
    ORS_OBJECT_IMP2(orsXImageLayer, orsIImageLayer, orsILayer,"default","Image
        Layer")
};
```

下面以 SetImageSource 来说明影像链的使用。影像的显示一般需要考虑影像的缩放、灰度拉伸等预处理功能,使最终用于显示的数据是亮度合适的字节类型。SetImageSource 根据影像的类型适当地增加处理环节。若是复数影像,则增加复数转模转换器;若需要放大,则增加影像放大器。为了进行影像的亮度调节,并把非字节影像转换为字节影像,则必须添加灰度拉伸器。

```
void orsXImageLayer::SetImageSource(orsIImageSource *pImgSource, bool bAutoZoom)
{
    if(pImgSource == NULL)
        return;

    //输入中已经包含自动放大器?
    if(ORS_PTR_CAST(orsIImageSourceWarper, pImgSource))
        bAutoZoom = true;

    orsIPlatform *pPlatform = getPlatform();
    m_imageSource = pImgSource;
    if(m_layerName.length()< 2)
        SetLayerName(orsString::getPureFileName(pImgSource->getFilePath()));
    //创建影像链
    m_imageChain = getImageService()->CreateImageChain();
    //复数到模转换器
    ref_ptr<orsIImageSource>complex2Mag;
    if(!bAutoZoom){
        //复数影像? 加上幅度计算节点
        if(pImgSource->getOutputDataType(0)> ORS_DT_FLOAT64)
            complex2Mag=ORS_CREATE_OBJECT(orsIImageSource,"ors.dataSource.im-
                age.complex2Magnitude");
    }
```

```
        //放大器
        ref_ptr<orsIImageSourceZoomer>  zoomer;
        if(!bAutoZoom)
            zoomer= ORS_CREATE_OBJECT(orsIImageSourceZoomer, ORS_IMAGESOURCE_ZOOM_
                DEFALUT);

        //拉伸器
        ref_ptr<orsIImageSource>imgMapper =
            ORS_CREATE_OBJECT(orsIImageSource, "ors.dataSource.image.mapper.2stdv");

        m_imageChain->add(pImgSource);
        //添加复数转换器
        if(complex2Mag.get())
            m_imageChain->add(complex2Mag.get());
        //添加灰度拉伸器
        if(imgMapper.get())
            m_imageChain->add(imgMapper.get());
        //添加放大器
        if(NULL != zoomer.get()){
            zoomer->setResampleMode(ORS_rsmNEAREST);
            m_imageChain->add(zoomer.get());
        }
        ...
        //初始化渲染引擎
        initImageRender();
        m_bVisible = true;
    }
```

2. 影像层渲染器

影像层渲染器 orsIImageRender 继承自渲染器 orsIRender。渲染器 orsIRender 定义的唯一的接口函数是显示参数设置 SetupParameterDlg,用于调用渲染器内部的对话框,进行显示参数的设置。

影像层渲染器分为单波段影像渲染器 orsISingleBandImageRender 和多波段影像渲染器 orsIMultiBandImageRender。影像渲染器,除渲染影像外,很重要的一个功能是图层树上显示风格的渲染。

```
    interface orsIRender : public orsIObject
```

```
{
    //显示参数设置
    virtual bool SetupParameterDlg() = 0;
    ORS_INTERFACE_DEF(orsIObject, "render");
};

interface orsIImageRender : public orsIRender
{
public:
    //设置图层,用于刷新图层显示风格更新通知
    virtual void setLayer(orsILayer * pLayer) = 0;
    //获取渲染器描述
    virtual orsString getRenderDesc() = 0;
    //设置要显示的影像链
    virtual void setImageChain(orsIImageChain * pImgChain) = 0;
    //渲染图层树上的显示风格
    virtual void rendImageLayerTree(orsHDC hDc, orsRect_i &dcrect, int flag) = 0;

    //影像区域显示
     virtual bool onDraw(BITMAP &dcBmp, orsImage2View * map2View, orsIRegion *
        pClipRgn) = 0;

    //获取图层树上的显示风格项,用于分配显示风格的绘制空间大小和文字
    virtual orsLayerStyleITEM * getStyleItemsOnTreeCtrl(int &nItems) = 0;
    ...

public:
    ORS_INTERFACE_DEF(orsIRender, "image");
};
```

　　如图 6-15 所示,单波段影像渲染借鉴 ArcMap 的思路,显示的模式包括彩色拉伸、Unique Value 显示、分类显示、RGB 组合和彩色晕渲等五类。彩色拉伸模式把影像灰度(0~255)映射为一个色条;Unique Value 显示为每个灰度级指定一个颜色,主要用于分类结果的显示;分类显示把灰度级分为若干个区间,并指定颜色进行显示;RGB 组合指定 R、G、B 三个分量是否显示,若某个分量不显示,则该分量将透明到下一个图层;彩色晕渲主要用于 DEM 的渲染。

```
    interface orsISingleBandImageRender : public orsIImageRender
    {
```

图 6-15 单波段影像层渲染对话框

```
//用于分类图的显示
public:
    //分类辅助文件的操作
    virtual orsString getClassInfoFileName() = 0; //获取辅助文件名
    virtual bool openClassInfoFile(orsString strFile) = 0; //打开辅助文件,若没有,
        //则建立默认分类
    virtual bool saveClassInfoFile(orsString strFile) = 0; //保持 Render 信息

    virtual COLORREF getClassRenderColorByDN(int dn) = 0;

    virtual int getNumOfClasses() = 0;
    virtual orsString getClassName(int ID) = 0;
    virtual bool setClassName(int ID, orsString strClsName) = 0;
    virtual COLORREF getClassRenderColorByID(int ID) = 0;
    virtual int getClassRenderSwitch(int ID) = 0;

    virtual bool setClassRenderColor(int ID, COLORREF colorRender) = 0;
    virtual bool setColorByClassID(int ID, BYTE r, BYTE g, BYTE b) = 0;
    virtual bool setClassRenderSwitch(int ID, int nswitch) = 0;

    virtual void createDefaultClassInfo(int defaultNum) = 0;

    //virtual void setClassRenderTable() = 0;
    virtual BYTE *getClassRenderTable() = 0;
```

```cpp
    virtual int getCapabilityOfClasses() = 0;
    virtual void setClassesNumChange(int numOfClasses) = 0;

//UniqueValue 显示
public:
    virtual void createUniqueValeClassInfo() = 0;

    virtual bool setUniqueValueColor(int ID, COLORREF colorRender) = 0;
    virtual bool setUniqueValueColor(int ID, BYTE r, BYTE g, BYTE b) = 0;
    virtual bool setUniqueValueSwitch(int ID, int nswitch) = 0;
    virtual bool setUniqueValueName(int ID, orsString strClsName) = 0;
    virtual int getNumOfUniqueValue() = 0;
    virtual int getUniqueValueSwitch(int ID) = 0;
    virtual COLORREF getUniqueValueColorByID(int ID) = 0;
    virtual COLORREF getUniqueValueColorByDN(int dn) = 0;
    virtual orsString getUniqueValueName(int ID) = 0;
    virtual BYTE * getUniqueValueTable() = 0;

//用于调色版,Stretch 的显示
public:
    virtual orsColorElements getColorElements() = 0;
    virtual void setColorElements(orsColorElements colorElements) = 0;
    virtual BYTE * getStretchRenderTable() = 0;
    virtual bool isStretchInvers() = 0;
    virtual void setStretchInvers(bool isInvers) = 0;
    virtual orsString getStretchTable() = 0;
    virtual void setStretchRender(orsString stretchTable) = 0;

public:
    ORS_INTERFACE_DEF(orsIImageRender, "singleBand");
};
```

多波段影像渲染主要提供波段组合功能,用于从多光谱或高光谱影像中选取三个波段作为 RGB 进行彩色显示。

```cpp
interface orsIMultiBandImageRender : public orsIImageRender
{
public:
    //获取显示的波段集
```

```cpp
    virtual orsBandSet getBandSet() = 0;
    //设置红色分量波段
    virtual void setBandR( int iBand) = 0;
    //设置绿色分量波段
    virtual void setBandG( int iBand) = 0;
    //设置蓝色分量波段
    virtual void setBandB( int iBand) = 0;
public:
    ORS_INTERFACE_DEF(orsIImageRender, "multiBand");
};
```

6.3.5 矢量图层及渲染

基于 Gdiplus 的线型和填充模式，实现简单要素类型的显示。

1. 简单要素矢量层

用于简单要素矢量层的渲染和简单要素的添加。

```cpp
class orsISFLayer : public orsILayer
{
public:
    virtual ~orsISFLayer(){};

    //添加简单要素
    virtual bool AppendFeature(OSF_wkbGeometryType type, orsPOINT3D * pts, int n)
        = 0;

public:
    //设置简单要素数据源层
    virtual void SetSFLayer(osfIVectorLayer * osfLayer) = 0;
    virtual osfIVectorLayer * GetSFLayer() = 0;

    //设置点渲染器
    virtual void SetPointRender(osfIPointRender * render) = 0;
    //设置线渲染器
    virtual void SetLineStringRender(osfILineStringRender * render) = 0;
    //设置面渲染器
    virtual void SetPolygonRender(osfIPolygonRender * render) = 0;
```

```cpp
public:
    virtual void DrawSelectedObject(orsHDC dc, orsRect_i &rect) = 0;

public:
    ORS_INTERFACE_DEF(orsILayer, "simpleFeature");
};
```

2. 矢量层渲染器

```cpp
interface orsISFRender : public orsIRender
{
public:
    //设置裁剪区域
    virtual void BeforeDraw(orsHDC hDC, orsIRegion * clipRgn = NULL) = 0;
    ORS_INTERFACE_DEF(orsIRender, "SF");
};

interface osfIPointRender : public orsISFRender
{
public:
    virtual void DrawPoint(orsPOINT2D * point, orsRenderMode rendMode) = 0;

public:
    ORS_INTERFACE_DEF(orsISFRender, "point");
};

interface osfILineStringRender : public orsISFRender
{
public:
    virtual void DrawLineString(orsPOINT2D * point, int n, orsRenderMode rendMode)
        = 0;

public:
    ORS_INTERFACE_DEF(orsISFRender, "lineString");
};

    interface osfIPolygonRender : public orsISFRender
{
```

```
public:
    virtual void DrawPolygon(orsPOINT2D * point, int * numOfP, int n, orsRenderMode
        rendMode) = 0;

public:
    ORS_INTERFACE_DEF(orsISFRender, "polygon");
};
```

6.4 综合显示与集成环境——orsViewer

6.4.1 设计目标

orsViewer 的设计目标是能够融合 ArcMap、ERDAS、ENVI 等主流 GIS、遥感软件的相关功能，形成一个具有较强栅格、矢量显示、能够动态集成影像处理功能的综合处理环境。主要功能包括以下几种。

1) 基本功能
(1) 通用图层管理。
(2) 快速漫游栅格矢量显示。
(3) 影像间假彩色波段组合。
(4) 分类图像彩色显示。
(5) 图层组卷帘。
(6) 矢量采集。
(7) 默认支持 GDAL 能读取的所有栅格格式。
(8) 默认支持 OGR 能读取的所有矢量格式。
(9) 内置基于 GdiPlus 的自定义线形、填充模式。

2) 界面扩展支持
(1) 支持自定义图层、图层组。
(2) 支持菜单、工具条、视图插入。

3) 按插件扩展实现的功能
(1) 矢量编辑。
(2) ROI 处理。
(3) 可执行对象菜单挂接(支持图层关联)。
(4) 嵌入式工作流(处理流)。

6.4.2 设计思路与界面设计

orsViewer 是一个以图层为核心的综合处理与显示程序，运用图层间的卷帘和假彩色合成实现输入、输出结果的对比。如图 6-16 所示，orsViewer 的界面主要由图层树、显示窗口、动态菜单及动态插入工具条、控件窗口等组成。加载的插件可以在菜单中增加菜单项，可以增加工具条，也可以直接插入整个窗口。图中的可执行对象树控制和嵌入式工作流就是动态插入的 BcgControlBar。

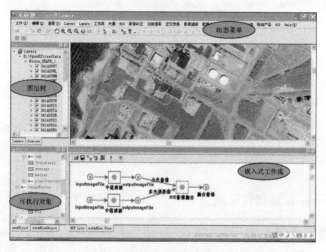

图 6-16　orsViewer 界面设计

6.4.3　orsViewer 扩展点设计

根据以上界面扩展需求，作者设计了 orsViewer 扩展点插件接口 orsIViewer-Extension。可动态扩展的界面元素包括菜单、工具条、视图、控制条；可动态扩展的图层内容包括自定义图层、影像渲染器、简单要素类型矢量渲染器等（图 6-17）。

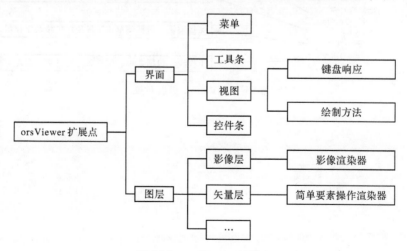

图 6-17　orsViewer 扩展点

orsViewer 扩展点插件接口 orsIViewerExtension 在 orsIGuiExtension 基础上增加了 orsViewer 的图层集合和影像显示接口，以便在扩展插件中对图层进行操作，并在显示窗口中进行交互和显示。orsIGuiExtension 代码如下：

```
interface orsIViewerExtension : public orsIGuiExtension
{
public:
    virtual void SetLayerCollection(orsILayerCollection *pLayerCollectioin) = 0;
    virtual void SetImageViewer(orsIImageView *pImageViewer) = 0;

public:
    ORS_INTERFACE_DEF(orsIGuiExtension, "orsViewer")
};
```

6.4.4 orsViewer 扩展插件示例

目前已经实现的 orsViewer 扩展插件包括 orsViewerExt_exeObject、orsViewerExt_roi、orsViewerExt_vecEdit、orsViewerExt_workflow 四种（图 6-18）。其中，orsViewerExt_exeObject 自动实现可执行对象的菜单挂接，orsViewerExt_roi 实现 orsViewer 的兴趣区选取，orsViewerExt_vecEdit 实现矢量图层的编辑、orsViewerExt_workflow 实现工作流的定义和执行。

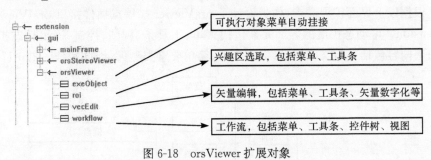

图 6-18　orsViewer 扩展对象

第7章 处理流程设计与实现

当前的遥感数据处理过程中，处理步骤越来越复杂，单独的处理算法已不能满足应用需求。需要对各种处理算法进行有效组合，构建合适的处理流程，从而生产出满足实际应用需要的遥感数据产品。

为了实现遥感数据处理流程的动态定制和处理任务的自动批处理，作者对各种处理流程进行综合分析，抽象出内存型处理流程和外存型处理流程两类，分别设计了两种不同类别的流程化处理模型，并在此基础上建立了可视化流程定制的处理框架。其中，内存型处理流程就是内部可连接对象，目前已经实现的主要是影像处理链 orsIImageChain，这部分内容已经在第4章阐述。本章主要介绍外存型处理流程。

7.1 外存型处理流程——可执行对象处理流

与内存型工作流不同，外存型处理流程的每一个处理节点执行时都涉及外存的输入输出，通过文件交换的方式将各个处理步骤串联起来。这种工作流处理模式在灵活性和适用范围等方面具有更大的优势，只需要定义好各个处理步骤的输入和输出，使得一个处理步骤的结果可以作为下一个处理步骤的输入，即可实现工作流的自动执行。通过共享存储作为交换中介，可执行对象和分布式并行处理相结合，可以实现局域网内的流程处理。在 OpenRS 中，作者设计了一种基于可执行对象的处理流程。

7.1.1 可执行对象

可执行对象是对数据处理算法的一种抽象，所有的处理算法都包括输入、输出和参数配置，而内部的处理过程相当于一个黑箱，由每个具体算法自己来实现。可执行对象制定了统一的输入、输出和参数配置接口，按照可执行对象的接口规则实现的算法插件对象，都可以针对这三类接口进行统一编程。如果将各个可执行对象的输入和输出串起来，即可实现各种处理算法的灵活组合，按需构造出不同的处理流程。

在可执行对象中，输入和输出一般比较简单，都是文件名，而参数配置较为复杂，对于不同的算法插件，参数个数、参数类型和参数的含义都不相同。为了对参数配置进行标准化，OpenRS 采用属性的方式实现参数的传递。采用属性方式的好处是可对不同的参数进行描述,可以对参数个数和参数类型进行灵活扩展,实现算法参数的动态定制。在参数传递的实现机制上,这种方法类似于 MFC 的序列化和逆序列化。

7.1.2 实现机制

可执行对象 orsIExecute 是由最初的 orsIObject 接口类派生的,在可执行对象的基础上分别派生出单机执行接口类 orsISimpleExe 和并行执行接口类 orsIParallelExe。这两个接口分别基于左面的单机执行方式和分布式并行执行方式。遥感影像处理算法只需要按照可执行对象实现 orsISimpleExe 接口或 orsIParallelExe 接口,即可以供桌面、分布式处理和网络服务调用。可执行对象的接口继承关系如图 7-1 所示。

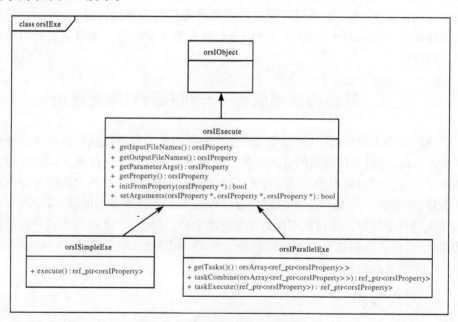

图 7-1　OpenRS 可执行对象接口图

下面以中值滤波为例,说明可执行对象的实现机制。中值滤波按单机执行和多机并行执行,分别实现了 orsSE_MedianFilter 和 orsPE_MedianFilter 插件对象(图 7-2)。orsSE_MedianFilter 对象按单机执行方式实现,在 execute 方法中实现了单机的中值滤波处理功能。orsPE_MedianFilter 按并行方式实现,需要实现 getTasks、taskExecute 和 taskCombine 三个方法。其中,getTasks 进行并行任务的分解,taskExecute 为任务的执行及中值滤波的具体实现,taskCombine 为任务的合并,具体原理可参看第 8 章。

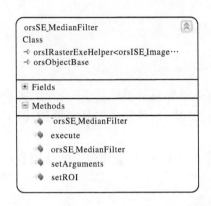 

图 7-2 两类中值滤波插件对象

按照统一的可执行对象接口，实现相应的方法，即可完成不同算法可执行对象插件的开发。OpenRS 中的可执行对象可以在不同的程序中调用，例如，直接在 orsExeRunner 中执行，或以菜单的形式挂到 orsViewer 的界面上执行。

7.2 基于构件模型的可视化流程定制

7.2.1 概述

工作流（Workflows）就是工作流程的计算模型（李红臣等，2003），其主要任务是将处理流程中的工作按照合适的逻辑和规则组织在一起，并以恰当的模型进行表示并对其实施计算。简单地说，工作流就是一系列相互衔接、自动执行的业务活动或任务。可以将整个业务过程看成一条河，其中流过的河水就是工作流。工作流能够实现任务的批处理，提高作业效率。一个完整的工作流处理模块应具备以下几个方面的功能：

（1）工作流模型的建立。
（2）工作流模型的装载与保存。
（3）工作流节点的解析。
（4）工作流的执行。

在 OpenRS 中，基于流程化处理模型设计了一种遥感数据工作流处理模块，实现了基于构件方式的可视化工作流定制及基于前置条件轮询方式的复杂流程处理模型。整个工作流处理模块主要包含以下几个部分：

（1）节点设计。
（2）节点关系。
（3）工作流模型读取与保存。

(4) 工作流执行。

下面对这四个方面的内容进行详细阐述。

7.2.2 节点设计

面向遥感数据处理的工作流,主要业务目标是遥感产品生产,为了实现这一目标,需要为遥感数据处理的参与者(各种算法)设计统一的计算模型,即工作节点(孙小涓等,2012)。

节点的设计是工作流设计的核心。在 OpenRS 中,工作流中的处理节点按如下方式设计:

```
struct orsWorkflowNode{
    int xPos,yPos;                          // 节点位置
    orsChar classId[80];                    // 节点 ID,用于实例化
    orsChar instanceId[80];                 // 实例名,用于标志实例
    int onlyMark;                           // 节点唯一标示 ID
    ref_ptr<orsIExecute>pObj;               // 本节点可执行对象
    orsArray<orsWorkflowNode * > pInputs;   // 输入节点
    orsArray<orsWorkflowNode * > pOutputs;  // 输出节点
private:
    int nPreConditions;                     // 前置条件计数
};
```

在节点模型中按照可执行对象方式,封装了遥感处理算法的实例,每个处理节点可以含有多个输入节点和输出节点,采用节点数组保存输入和输出。此外,节点模型中还包含一个前置条件计数 nPreConditions,可在工作流中控制当前节点是否可执行。当前置条件计数满足规定要求时,当前节点可被执行。

整个工作流便是由以上节点的集合组成的。通过一个节点数组来定义工作流中的全部节点。

```
orsArray<orsWorkflowNode * >m_pNodes;
```

7.2.3 节点关系

节点模型定义好以后,就可以开始建立工作流。在建立工作流时,首先需要添加节点对象,并根据节点 orsIExecute ID 初始化当前节点的可执行对象。然后就需要确认各节点的输入和输出,从而建立节点之间的关系。在 OpenRS 中,每个节点只包含输入和输出两种关系,一个节点可以具有多个输入,也可以具有多个输出。这

些输入和输出节点分别保存在当前节点的成员变量 pInputs 和 pOutputs 中。

根据节点输入和输出的数量,将节点分为三类:若一个节点对象只有输出节点,没有输入节点,则称为头节点;只有输入节点,没有输出节点,则称为尾节点;具有输入节点,也有输出节点,则称为中间节点。

各个节点的输入和输出个数是由节点内部处理算法及当前节点的可执行对象决定的。如图 7-3 所示,左侧为中值滤波节点,仅包含一个输入和一个输出,而右侧为 PCA 影像融合节点,包含两个输入和一个输出。

图 7-3　两种处理节点

节点执行时的参数是通过 OpenRS 中的属性 orsIProperty 进行配置的,可以通过属性节点对可执行对象进行参数配置。由于各节点的可执行对象参数都是用 orsIProperty 进行存储的,所以在设置时可以用统一的针对 orsIProperty 设计的交互界面 orsIGuiService 来配置,当然也可以使用用户自定义的界面 orsIExeConfigDlg。

7.2.4　工作流的序列化

完整的工作流建好后,可以将工作流模型进行保存,方便以后进行同类处理时直接调用。OpenRS 中提供了工作流的序列化机制,实现工作流的导入和导出。工作流模型按照 xml 格式进行保存,在保存工作流时,对所有的工作流节点进行编号,然后逐个保存各个节点。工作流导入时也是根据节点编号,将节点顺序导入工作流的节点数组中。xml 文件的读写可以使用 OpenRS 内部的属性序列化插件 orsIPropertySerialize 实现。

在节点运行前,还必须对节点进行解析。解析工作主要是为了确定该节点运行所需的依赖条件。在工作流中,一个节点的运行需要依赖于输入节点的运行结果,而其所依赖的节点又需要依赖另一些节点。采用递归方式对节点的依赖关系进行解析,并重建工作流中的节点关系。每个节点在序列化时都记录了其输入节点和输出节点编号,在读取时依据其记录的输入输出节点编号,将各个处理节点连接起来。OpenRS 的工作流处理模块中专门提供了一个函数 buildNodeLinkage 来重建节点之间的输入输出关系。

7.2.5　工作流的执行

工作流中各个处理节点间的执行是有逻辑顺序的(高明,2013),OpenRS 中的工作流执行时采用类似令牌的方式,每次只有满足前置条件的节点才可以被执行。若不是在并行处理环境中,则每次只有一个节点被执行。对于一个节点,若全部前置条件得到满足,则当前节点可以执行,并产生相应输出。若前置条件不满足,则当前节点处于等待状态,等待其他节点先执行。当前节点执行完后,需要更新该节点输出节点的前置条件。工作流执行器不断地查询每个输入节点的前置条件来推进工作流的执行,直到所有节点都被执行。

7.2.6 OpenRS 中的工作流插件

OpenRS 中的工作流处理模块采用界面扩展插件的方式集成到遥感综合处理模块 orsViewer 中。图 7-4 为嵌入 orsViewer 中的工作流处理模块界面。左侧为可执行对象插件树,右下侧为工作流窗口。可以通过简单的"拖拽"操作将所需要的可执行对象拉到工作流窗口中,可视化地定制工作流。将不同可执行对象的输入和输出拖到一起,工作流管理器会自动将输入节点和输出节点连接起来,构成完整的流程。

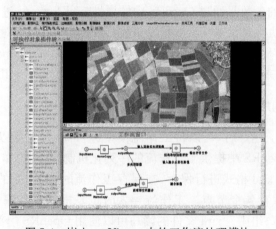

图 7-4 嵌入 orsViewer 中的工作流处理模块

7.3 面向处理流程的分布式批处理

工作流批处理模块结合了工作流和批处理的优势,实现流程的批量处理功能。结合 OpenRS 平台中的并行处理模式,工作流批处理模块能够根据指定的工作流文件,对相应的遥感数据进行并行处理,满足了不同类型数据的流程定制和批量处理的要求。有关分布式并行处理请参看第 8 章。

7.3.1 原理

OpenRS 平台上,工作流批处理模块的实现主要依赖可视化流程定制、PTR

并行处理模式(见第8章)和分布式文件系统三个方面。其中,分布式文件系统只是一个性能选项,本书不展开讨论。

首先,可视化流程定制实现了基本算法的组装和集成,提供了根据具体业务需求定制数据处理流程的基本功能。可视化流程定制的详细介绍参见7.2节。

其次,PTR建立一个支持快速、高效的并行处理,同时功能可扩展的遥感数据处理服务。在已有高效网络通信中间件(internet communications engine,ICE)的基础上,PTR可以利用廉价的PC群进行高性能遥感数据处理,最终提供网络透明,扩展性强,多粒度的统一并行计算框架,供OpenRS项目中各种算法集成与处理。因此,PTR并行处理为基于工作流的批处理模块提供了轻量级的多任务并行处理的环境。

最后,PFS(parallel file system)是一个轻量级分布式文件系统,主要面向存储密集型遥感数据处理,通过计算机网络将遥感数据文件存储在不同的网络节点,提供多个磁盘节点的并发访问来获取比本地磁盘更优的聚合IO吞吐量。系统设计基于客户机/服务器模式,逻辑组成主要由应用程序、操作系统内核、文件访问驱动、文件目录索引服务和文件存储服务组成,通过虚拟化技术为应用程序提供类似本地驱动器的访问方式,应用程序不需做任何修改。

7.3.2 实现

OpenRS的工作流批处理模块主要由工作流选择区、数据选择区、工作流编辑区和任务管理区组成。工作流批处理模块的界面组成如图7-5所示。

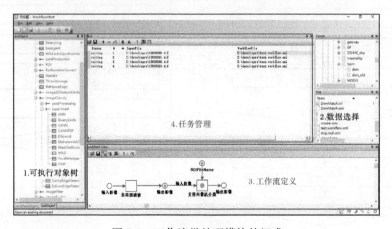

图7-5 工作流批处理模块的组成

其中,工作流选择区域显示了OpenRS平台工作流空间(WorkflowSpace)下所有配置好的工作流文件。用户可以将需要使用的工作流文件拖放到任务管理区域,为指定的数据配置相应的工作流文件。同时,工作流选择区域中的文件也可以拖放到工作流编辑区域，对工作流文件进行修改、保存或者单任务执行。

数据选择区域用于将需要处理的影像数据配置到任务管理区域。数据管理区域实现了单个影像文件，多个影像文件与指定工作流文件的拖放，即用户可以选择单个影像文件和单个工作流文件，多个影像文件和单个工作流文件的搭配，或者仅配置任务管理区域没有指定工作流文件的影像文件。

任务管理区域显示了用户配置的所有任务信息，如任务状态，处理文件和对应工作流文件等。任务管理区域实现了任务的打开、保存、增加任务、删除任务、执行批处理和系统配置等功能。任务管理区域是工作流批处理最重要的组成部分。

根据以上四个区域的协同配合，OpenRS 平台中的基于工作流批处理模块实现了配置工作流、修改工作流、工作流管理、任务配置和任务批量并行处理等功能。

7.3.3 执行

批处理的执行是以工作流节点为单位的，用专门的一个线程来实现节点任务的提交和状态查询。批处理程序每次只提交已经准备好的、可以执行的节点任务。同时执行的节点任务可以同时提交给分布式并行任务管理器，由并行任务管理器负责分配执行的硬件资源。而批处理管理器只负责检查每个流程任务节点的执行状态，并提交新的任务。具体批处理执行线程的流程如图 7-6 所示。

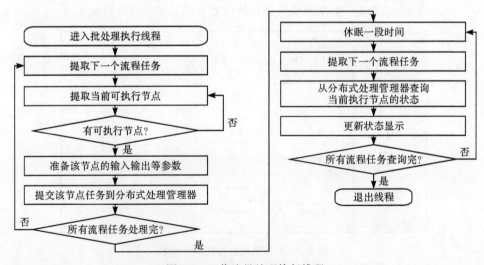

图 7-6　工作流批处理执行线程

第8章 分布式并行处理环境设计与实现

目前遥感和对地观测技术经过多年的发展,取得了巨大的进步,并随着我国卫星自主研发能力的提高,即将发射一系列高分辨率遥感对地观测卫星,建立覆盖可见光、红外、多光谱、超光谱、微波、激光等观测谱段的、高中低轨道结合的、具有全天时、全天候、全球观测能力的大气、陆地、海洋先进观测体系,以提高我国空间数据的自给率,形成空间信息产业链条。在这样的环境和要求下,对于遥感影像处理的研究不能仅限于算法本身的研究,还需要综合考虑采用何种模式和技术手段将未来遥感数据处理水平在服务模式、海量化处理、开发方式、商业模式、技术与资源共享、系统可持续性等方面提升一个新的水平。

分布式并行处理作为最重要的并行处理模式之一,一直是高性能处理的主流(祝若鑫等,2015)。虽然 GPU 加速发展,使并行加速从几个十几个的 CPU 多核并行发展到数以百计的 GPU 并行,但多机加速必然会发展到多机 GPU 加速,分布式并行处理仍将是一种不过时的高性能处理模式。

本章基于 Google 公司提出的 MapReduce 编程模型(Google MapReduce 中文版),介绍自主设计实现的轻量级分布式并行处理系统(parallel task runner,PTR)。目前,基于 PTR 的 OpenRS 分布式服务处理系统已经作为 OpenRS-Cloud 的后台核心应用于遥感云服务在线处理系统。

8.1 PTR 并行模型与功能特性

8.1.1 MapReduce

Hadoop Map/Reduce 是一个使用简易的软件框架,基于它写出来的应用程序能够运行在由上千个商用机器组成的大型集群上,并以一种可靠容错的方式并行处理上 T 级别的数据集(彭辅权,2012)。其中,MapReduce 是 Google 提出的一种用于大规模数据集的并行处理模型,其主要思想来源于函数式编程语言和矢量编程语言。MapReduce 模型主要由"Map(映射)"和"Reduce(化简)"组成。Map 函数的主要功能是将原始输入映射成〈Key,Value〉数据。映射操作是可以高度并行的,这对高性能要求的应用及并行计算领域的需求非常有用。处理系统将映射操作输出的〈Key,Value〉根据 Key 值发达给相应的 Reduce 函数。Reduce 函数对接收到的〈Key,Value〉列表进行处理,一般会输出一个新的由〈Key,Value〉表达的较小的独立元素列表。通常作业的输入和输出都会存储在文件系统中。整个框架负责任务的调度和监控,并重新执行已经失败的任务(图8-1)。

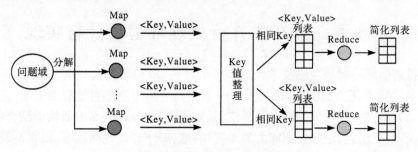

图 8-1　MapReduce 并行模型

MapReduce 计算模型已经被应用在了如 Google App 和 Hadoop 的著名云计算平台(Golpayegani et al.,2009)。Apache 的 Mahout 项目基于 MapReduce 模式并行实现了较多平行计算环境下的机器学习算法。尽管 MapReduce 模式具有良好的可伸缩性,但对于大部分的遥感处理算法来说,只需任务分解、任务执行和结果合并三个步骤,不需要〈Key,Value〉相关概念,并且 MapReduce 模式产生的"Value"是处理后的结果数据,而遥感影像处理的结果数据是一个范围内的影像,如果将这个影像直接作为"Value"放入输出中,将极大影响数据处理的效率。Hadoop 的原生语言支持为 Java,C++语言程序可采用 Hadoop Streaming 和 Hadoop Pipes 进行支持,但在执行反馈、调用方式等方面有待提高。另外,Hadoop 对更复杂的调度(如任务之间存在多个层次依赖)支持不好,且调度时间花费较长,平台部署复杂。OpenRS 中所有算法均采用 C++编写,使用 Hadoop 存在局限。

除上述运行于计算集群上的并行系统外,还有许多适应于单机多核并行环境的软件系统。例如:以 C++模板库形式,跨平台的 Intel thread building blocks (TBB);支持跨平台共享内存方式的多线程并发的编程语言(open multi-processing,OpenMP);由 NVIDIA 所推出,适用于 GPU 并行处理的(compute unified device architecture,CUDA)语言;适合 CPU、GPU 或其他类型处理器组成的异构平台的并行编程语言及框架(open computing language,OpenCL)。

综上所述,未来大规模并行处理技术的发展是单机加速和分布式并行处理的结合。OpenRS 架构支持分布式并行处理,而每个处理节点上的多核加速或 GPU 加速由算法自身实现。OpenRS 的并行系统 PTR 是在 Google MapReduce 模型基础上针对遥感数据处理的特点扩展而成的,能够支持更加复杂的计算模型。

8.1.2 功能特性

OpenRS 并行系统 PTR 的目标是建立一个适合遥感影像并行处理,具备高效、灵活、轻量、部署简单、功能可扩展等特性的新型并行处理框架。PTR 提供多层次并行抽象模型,支持子任务级并行算法、任务及服务状态监控,提供本地模拟调试环境,可与 OpenRS 插件系统无缝集成。表 8-1 是 PTR 与现有通用领域并行处理系统的特性对比。

表 8-1 PTR 特性对比分析

特性	Condor	MPI	Hadoop	PTR
并行模型	不提供	不提供	提供	提供
算法网络透明	透明	不透明	透明	透明
支持语言	任何	C++、Java 等	Java,C++ 部分支持	C++
系统容错能力	提供	不提供	提供	提供
状态监控	任务日志	消息传递	直接提供	直接提供
系统部署	简单	较复杂	复杂	简单
并行粒度	任务级	精细粒度	子任务级	子任务级
模拟调试环境	不提供	不提供	提供	提供

PTR 针对如何对各种遥感处理进行统一的高性能服务实现问题,在分析遥感处理并行特点和现有并行模型优缺点的基础上,提出了面向兴趣区(ROI-oriented)、分解/合并(Decompose/Merge)、层次子任务链三个层次的多粒度并行模型。对于平台软件,系统可将该三个粒度的并行接口转换成统一的调度和容错流程;对于算法编写者,可根据不同的算法并行特点选择不同实现难易程度的并行抽象接口模型进行实现。下面将介绍 PTR 并行处理框架的概念角色组成、基本的计算资源调度原理及并行算法提交与状态监控实例。并行抽象接口将在 8.3 节详细阐述。

8.2 PTR 并行处理框架

8.2.1 系统角色组成

PTR 在上述遥感处理角色中主要扮演位于中心的平台架构者角色,是算法提供者与最终用户的桥梁。从物理部署角度,PTR 可分为调度服务节点、计算节点、任务提交、状态监控、分布式文件存储节点与 OpenRS 插件系统六个角色。其中,调度服务节点全局唯一,提供任务执行过程和机器的监控、任务提交、资源调度等工作;计算节点可动态加入调度服务节点提供计算资源;任务提交与状态监控均为客户端,由最终用户调用;OpenRS 插件系统负责提供各类算法插件本地调用执行,各类算法在经过算法提供者编写完成后,放入预先部署在计算节点的插件系统中;分布式文件存储节点主要负责提供统一、透明的分布式文件系统,供计算节点间数据交换、输入输出存储等,参见图 8-2。

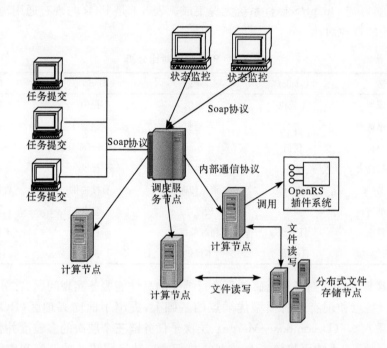

图 8-2　遥感并行处理系统物理部署图

调度服务节点和客户端均通过标准 Soap 协议进行通信;调度服务节点和计算节点之间利用内部高效协议通信;分布式文件系统提供标准的 POSIX(portable operating system interface)文件访问接口。下面对每个角色进行详细介绍,并给出相关软件程序包。

1. 调度服务节点

任务管理与调度接收来自 Web 服务网关的各种任务监视、执行等请求,负责给任务分配计算资源。其内部维护一个任务优先级队列和可用计算资源列表,是客户端(任务提交与状态监控)远程调用的唯一入口。通过监视远程计算资源(任务接收与执行模块)的运行状态,合理进行任务级分配和管理,但不进行子任务管理,子任务管理由任务接收与执行模块内部完成。为了保证系统稳定,任务管理与调度并不执行任何外部算法,而且当某个具体任务执行完毕后,对相关利用过的计算资源进行清理(远程进程重启)以避免内存泄漏等问题。任务提交时可以设定任务的优先级,调度节点内置优先级队列,任务根据优先级依次启动。

调度服务节点的程序形态为可执行程序 ManagerServerd.exe,位于\OpenRs\ptr\PTRService 目录,在对其进行配置(配置文件:ptr.cfg)后,可直接单击运行。相关配置信息参见附录 C.2。

2. 计算节点

任务的具体执行由计算节点完成,调度服务节点为根据资源空闲情况分配给每个任务若干计算节点。这些计算节点可分为两类,分别为任务调度者和任务执行者,每个运行任务接收与执行的远程进程均可以扮演这两种角色,角色由调度服务节点指定。任务调度者由一个计算节点承担,根据并行模型(参见 8.3 节)将原始任务分解为多个子任务,然后将子任务发送给任务的其他计算节点进行并行处理;充当任务执行者的计算节点群主要是负责子任务的执行,由插件执行代理调用具体算法完成。插件执行代理位于任务接收与执行模块和具体的算法插件之间,负责从任务信息和子任务信息中抽取出算法 ID 和相关输入,然后调用算法插件系统,进行具体的算法执行过程,并负责将算法执行反馈信息(进度信息)传递给任务接收与执行。

计算节点由 WorkServer 与 PTRAgent 两个程序组成。其中,WorkServer.exe(位于\OpenRs\ptr\PTRService 目录)负责网络服务,相关配置信息参见附录 C.2;PTRAgent(位于\OpenRs\desktop\debug\vc60)负责插件执行代理,必须位于 OpenRS 主目录。

3. 客户端(任务提交与状态监控)

客户端主要包括两个部分的内容,分别为任务提交与状态监控。客户端的软件形态有三种,其一为 C++调用封装类包(orsParallel.dll),主要提供 C++接口供第三方软件二次开发调用,提供任务提交与受限的状态监控功能;其二为标准 Webservice 服务接口,在启动 PTR 的 ManagerServerd 后,可通过 http://localhost:18083/? wsdl 获取服务的 wsdl 描述信息,便于非 C++语言调用,提供任务调用与完整的状态监控功能;其三在启动 PTR 的 ManagerServerd 后,监控用户可直接访问 http://localhost:18083/ptr/taskManager.html,获取交互式监控界面,主要面向系统用户,支持完整的状态监控功能。

4. 分布式文件存储节点

PTR 系统提供一个轻量级分布式文件系统(parallel file system,PFS),主要面向存储密集型遥感数据处理,通过计算机网络将遥感数据文件存储在不同的网络节点,提供多个磁盘节点的并发访问来获取比本地磁盘更优的聚合 IO 吞吐量。PTR 并不绑定 PFS,用户可根据需要选择其他成熟分布式文件系统。

PFS 系统设计基于客户机/服务器模式,主要由应用程序、操作系统内核、文件访问驱动、文件目录索引服务和文件存储服务组成,通过虚拟化技术为应用程序提供类似本地驱动器的访问方式,应用程序不需做任何修改。参见图 8-3。其中,应用程序通过标准操作系统 API 向操作系统内核发起文件访问请求,操作系统内核将请求转发给 PFS 文件访问驱动,驱动首先通过网络远程访问文件目录索引服务来获取具体的文件存储网络位置,然后通过该网络位置与具体的文件存储服务直接连接,并获取具体的文件数据,然后返回给应用程序。文件目录索引服务主要存储了文件的目录结构信息和具体的文件网络索引信息;文件存储服务则记录具体的文件数据实体信息;文件访问驱动主要起与操作系统内核衔接的作用。

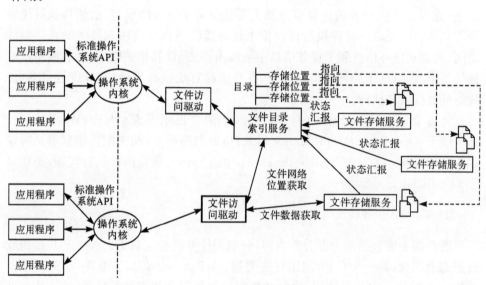

图 8-3 PFS 分布式文件系统逻辑组成

8.2.2 并行调度原理

PTR 的并行调度分为两个层次,分别为计算资源调度与任务执行调度。其中,计算资源调度发生在调度服务节点上,某个任务执行前,调度服务节点启动计算资源调度流程为其分配合适的空闲计算节点组,供其在后续任务执行调度中的并行子任务处理;任务执行调度发生在计算节点上,主要是控制子任务的分解与合并过程,并负责将子任务推送到合适的计算节点上执行。下面,将分别进行阐述。

1. 计算资源调度

基于进程池的高可用资源调度框架是 PTR 计算资源调度核心,通过将运行在分布式环境下的各个具体计算节点映射为驻守在"调度服务节点"服务器上的虚拟资源,并在任务请求时,将这些资源合理分配给计算节点上的"任务调度者",任务调度者内部形成远程进程池,不断地推送相关子任务到计算节点上的"任务执行者"远程进程中执行,并获取反馈信息,参见图 8-4。

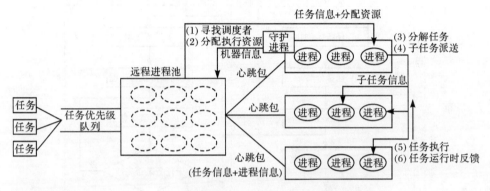

图 8-4 基于进程池的高可用资源调度框架原理图

为了保持系统的稳定,计算节点以心跳包的形式,将自身服务器资源空闲状况(包括硬件信息与执行信息)发送给中心调度服务节点。该功能一方面为了状态监控使用,另一方面为后续任务执行调度使用。由于任务执行服务器众多,可能会出现硬件失效的情况,人工查看硬件成本太高,不利于多机管理,该状态汇报可以实现系统的自动诊断,方便管理人员,也可以实现资源的负载平衡。如果调度服务节点的任务队列中有未完成的作业,且有合适数量的空闲计算节点,那么将启动任务资源分配过程,即从空闲计算节点组中选择一个任务调度节点和若干任务执行节点,然后将任务信息推送给任务调度节点,由任务调度节点启动任务执行调度过程。另外,计算资源调度根据子任务进度状态,对子任务执行进程的状态和生命周期进行管理,当任务执行完成或失败时,进行远程进程清理,避免内存泄漏。

2. 任务执行调度

任务执行调度过程在任务调度计算节点上执行。任务执行调度过程主要完成两个内容,一是任务分解与发送;二是反馈子任务执行情况给调度服务节点。任务分解与发送主要基于一组 Dynamic Task Mode 接口,该组接口是 PTR 并行抽象接口模型(参见 8.4 节)的基础支撑,各类并行抽象接口模型最终均是由符合

Dynamic Task Mode 接口的代理程序(PTRAgent)代为执行的。PTRAgent 也是算法插件与计算节点服务(WorkServer)之间的信息传递桥梁,对子任务的执行情况(包括任务状态、执行进度等)进行快速反馈。

任务调度根据并行模型将原始任务分解为多个子任务,然后将子任务发送给任务执行者进行并行处理;任务执行通过调用插件执行代理来进行子任务的具体处理。参见图 8-5。

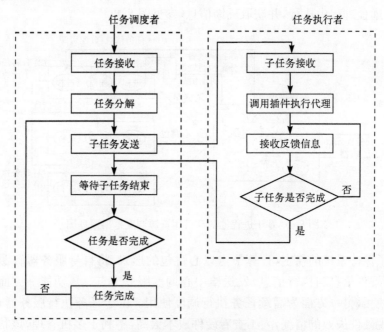

图 8-5　任务接收与处理流程图

Dynamic Task Mode 作为整个模型的底层,覆盖算法最广,也最灵活,虽然用户也可以自行实现,但相比 PTR 并行抽象接口而言,编写难度较大,实现的抽象接口较多。因此,OpenRS 提供了默认实现软件包——PTRAgent。该软件包主要包括 Job 与 Task 接口,Task 代表子任务,Job 代表整体任务。Dynamic Task Mode 接口模型允许在计算过程中产生新动态子任务,无需固定子任务之间的拓扑依赖关系。接口具体描述如下:

```
struct AlgTaskCallback
{
    virtual void onProgress(float percent) = 0;
    virtual void onFailed(char * why) = 0;
};
```

```cpp
struct AlgTask
{
    //TaskID
    virtual const char * getID() = 0;
    //Task 的类型
    virtual int getWorkType() = 0;
    //预处理,在发送到任务执行节点前,在中心节点被执行
    //可自己提取相关文件块数据
    virtual int preProcess() = 0;
    //序列化,在中心节点执行,任务将自身信息进行序列化
    virtual int getInputParams(AlgBlob* inputParam) = 0;
};

struct AlgJob
{
    //得到并行数
    virtual int getMaxPNum() = 0;
    //得到 Job 工作类型
    virtual int getJobWorkType() = 0;
    //不断得到子任务,直到为 NULL,在中心节点执行
    virtual AlgTask * GetTask() = 0;

    //当子任务完成时,被调用,在中心节点执行
    //自己管理任务的生命,看是否被释放
    virtual int onTaskFinish(AlgTask *cTask,AlgBlob* outputParam) = 0;
    //判断 Job 是否结束
    virtual bool isFinish() = 0;

    //得到大概的所有任务个数,用于后续进度百分比估计
    virtual int getMaybeAllTaskNum() = 0;
};

//PTRAgent 插件执行代理实现的相关接口
typedef void* (*initAgentFun)();
//根据 Job 类型,创建对应的 Job 算法实例
typedef AlgJob * (*createAlgByTypeFun)(const char * jobType,const char * algID,
    AlgBlob * jobInput,int resNum);
typedef void (*freeAlgFun)(AlgJob* job);
```

```
//具体子任务执行函数
//执行处理,并返回结果,inputParam 由 sendToExecute 得到
//callback 可能为 NULL
typedef void * AlgTaskHandle;


typedef AlgTaskHandle ( * createAlgTaskExeFun)(const char * jobType, const char *
    jobID, const char * algID, AlgBlob * jobInput);
typedef int ( * doAlgTaskFun) (AlgTaskHandle handle, AlgMsgSender * sender,
    AlgMsgRecver * recver, AlgBlob * inputParam, AlgBlob * outputParam, AlgTaskCall-
    back * callback);
typedef void ( * destoryAlgTaskHandleFun)(AlgTaskHandle handle);
typedef const char * ( * agentErrorInfoFun)();
```

基于上述并行模型接口,设计了 PTR 的并行调度算法,具体算法流程参见图 8-6。在调度算法中,taskArray 是由并行调度平台提供的执行任务队列,主要有 Push、PopFinishTask 与 WaitForTaskFinish 三个主要函数。Push 用于加入待执行的 task;当有 task 完成时,taskArray 会保存在内部完成队列中,由 PopFinishTask 弹出返回;WaitForTaskFinish 用于阻塞(block)等待内部完成队列,当内部完成队列不为空时,阻塞解除。另外,并行调度平台还提供函数 TaskDispatchAsyn,以非阻塞方式将 task 加入执行队列等待执行。

良好的容错机制是检验并行计算模型是否有效的标准之一。对于上述调度算法,需要考虑任务算法内部逻辑出错与子任务执行出错两种情况。上述任务调度逻辑并不在中心调度服务器上运行,而是发送给某个空闲服务代为执行,因此任务算法内部逻辑出错不会影响中心调度服务器而导致整个系统崩溃。但由于该类错误涉及整个任务,因此出错时,需根据任务的输入全部重做,以完成的子任务废弃。

```
Algorithm:Dynamic Task Mode Parallel Dispatch
Process Input:Job
Output:Job Process Result
stop:=false; tasknum:=0; jobfailed:=false;
taskArray.Push(inittask);
while(not stop)
    taskFinish=taskArray.PopFinishTask
    if(taskFinish not empty)
        if(taskFinish.Status is error)
            stop:=true;
            jobfailed:=true;
        else
            tasknum--;
            job.OnTaskFinish(taskFinish);
        endif
    else
        if(job.IsFinish()and tasknum==0)
            stop:=true;
        else
            task=job.GetTask();
            if(task is empty)
                if(tasknum==0)
                    stop:=true;
                else
                    taskArray.WaitForTask-
                        Finish();
                endif
            else
                tasknum++;
                TaskDispatchAsyn(task.Do
                    (job));
            endif
        endif
    endif
endwhile
```

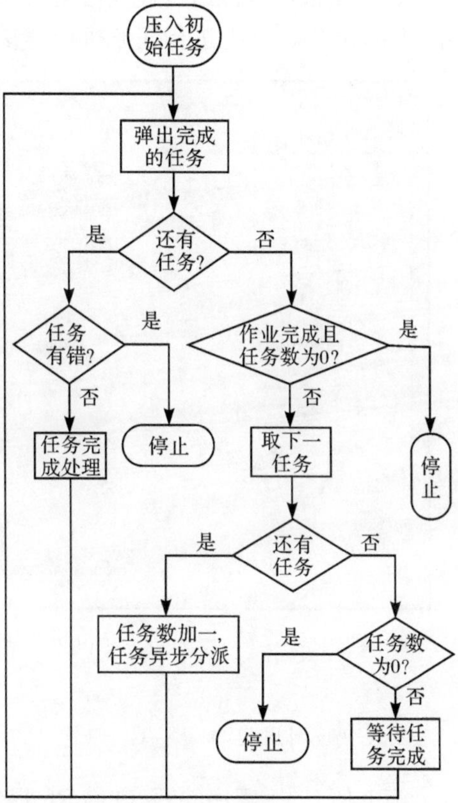

图 8-6 基于进程池的高可用资源调度框架原理图

8.2.3 并行算法提交

OpenRS 算法有三种入口方式可以提交给调度服务节点进行并行执行,分别为界面交互式提交、C++接口及标准 WebService 接口。下面分别进行介绍。

1. 界面交互式提交

界面交互式提交采用 OpenRS 提供的 orsExeRunner.exe 程序。用户在配置文件 openRS_Parall.conf 中的 MODE：PTR，URL：http：//localhost：18083/后，即可启动 orsExeRunner.exe，选择算法，填写算法输入、输出及算法参数后，选择 nProcess 个数（并行数，也代表占用的远程计算资源），单击 Run Remote 提交到远程 ManagerServer 上进行任务执行。参见图 8-7。

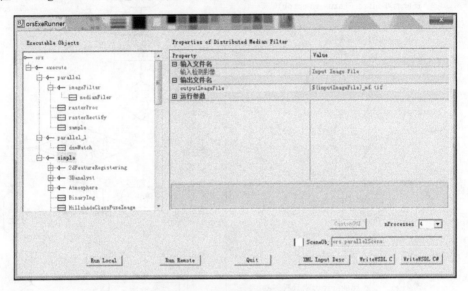

图 8-7 界面交互式提交

2. C++ 调用接口

为方便 C++ 程序编写，OpenRS 提供了 orsIParallelService 接口，其服务唯一标识为"ors.service.parallel"，接口形式如下。

```
interface orsIParallelService : public orsIService
{
public:
    //得到远程计算机群相关信息
    virtual  bool getMachineInfos(orsArray〈orsMachineInfo〉&machineInfos)=0;
    //得到可用的远程计算资源数
    virtualint getAvailableProcesses()=0;

    //任务提交(无进度返回)
```

```
//serverUrl:管理调度节点 url,如:http://localhost:18083/
//usr:提交用户名
//password:用户密码
//prallelExeClass:算法类唯一ID
//jobParas:任务参数
virtual bool runJob_parallel(const orsChar * serverUrl, const orsChar * user,
    const orsChar * password, const orsChar * prallelExeClass, orsIProperty *
    jobParas)=0;

//本地模拟提交(无进度返回)
virtual bool runJob_local(const orsChar * prallelExeClass, orsIProperty * job-
    Paras) = 0;
//任务提交(返回进度)
virtual bool runJob_parallel(const orsChar * serverUrl, const orsChar * user,
    const orsChar * password, const orsChar * prallelExeClass, orsIProperty *
    jobParas, orsIProcessMsgBar3 * pMsgBar)=0;
//本地模拟提交(返回进度)
virtual bool runJob_local(const orsChar * prallelExeClass, orsIProperty * job-
    Paras, orsIProcessMsgBar3 * pMsgBar)=0;

    ORS_INTERFACE_DEF(orsIService, "parallel");
};
```

orsIParallelService 一般可采用如下模板式代码进行调用:

```
//exeObj 为可执行对象(继承 orsIExecute),外部创建
void OnRunRemote(orsIExecute * exeObj)
{
    //可执行对象的参数(输入文件、输出文件及算法参数)
    orsIProperty * jobParas = exeObj->getProperty();
    //获取并行服务对象
    orsIParallelService * pService = getParallelService();

    //创建进度界面对象
    ref_ptr<orsIProcessMsgBar3> progressBar = ORS_CREATE_OBJECT(orsI-
        ProcessMsgBar3, ORS_PROCESSMSG_BAR3_DEFAULT);

    //得到可执行对象的唯一 ID
    orsString objID = exeObj->getClassID();
```

```
            //如果可执行对象是简单可执行对象
            if (ORS_PTR_CAST(orsISimpleExe, exeObj)){
                //自动并行
                if (m_nProcesses > 1){
                    //要被自动化并行执行的对象
                    jobParas->addAttr(ORS_RASTER_EXE_OBJECT, exeObj->getClassID(),
                        true);
                    //取自动并行化对象
                    orsString peObjID;
                    jobParas->getAttr(ORS_PARALLELIZE_OBJECT, objID);
                }
            }
            jobParas->addAttr(ORS_PARALLEL_NUM_PROCESSES, m_nProcesses, true);

            if (!pService->runJob_parallel(NULL, "jws", "jws", objID, jobParas, progress-
                Bar.get()))
            {
                printf("Job Failed");
                return ;
            }
        }
```

3. 标准 Webservice 网络服务接口

遥感云服务 Web 前端界面一般采用 Flex、JavaScript 等语言编写,为兼容这类语言,OpenRS 提供了标准 Soap 协议的 Webservice 接口,不同编程语言可通过 http://localhost:18083/?wsdl 获取该 Webservice 的 WSDL 描述,继而生成相关的代理代码。下面是 C#语言形式的代理代码。

```
public class PTRService : System.Web.Services.WebService
{
    //username:用户名
    //jobType:保留
    //jobInfo:保留
    //algID:算法 ID
    //inputParams:输入参数,orsIProperty 序列化的 XML 字符串
    //priority:优先级
    //timeOut:指定算法超时时间,单位为毫秒,若在 timeOut 间隔内算法未输出进度
```

//信息,则认为算法失败
```
int submitJob(string username, string jobType, string jobInfo, string algID,
    string inputParams, int priority, int timeOut, out string jobID)
}
```

8.2.4 执行状态监控

OpenRS 任务状态监控提供了两种主要监控形式,一种是网页型监控客户端,另一种是标准 Soap 协议的 Webservice 接口。下面分别进行阐述。

1. 网页型监控客户端

网页型监控客户端主要是利用 Flex 富客户端技术,以 Soap 协议为通信手段,通过与 Web 服务网关交互,实现任务监控、机器监控等的信息监视。用户在启动 ManagerServer 服务程序后,可直接输入 http://localhost:18083/ptr/taskManager.html,填写正确的 ManagerServer 的 IP 地址后,获取监控信息。参见图 8-8。

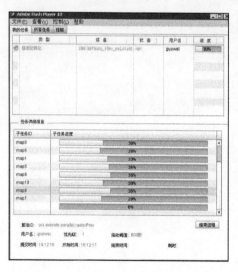

图 8-8 网页型监控客户端界面

网页型监控客户端主要包括如下功能。

(1) 个人任务管理:管理个人提交的任务(任务属性包括类型、信息、任务状态、用户名和任务的提交进度)。每个任务还对应了其任务信息(算法 ID、用户名、优先级、活动阈值、提交时间、开始时间、结束时间和耗时)和子任务信息(子任务 ID、子任务)。

(2) 所有任务管理:管理所有用户提交的任务信息(包括类型、信息、状态、用户名、任务进度)及任务进程的关闭管理。

(3) 计算服务器管理:管理当前运行所有计算机的状况,具体包括机器名、机器核数、内存、CPU、硬盘的使用和计算机各个进程运行的状况。

2. WebService 状态监控接口

WebService 状态监控接口基于 Soap 协议,不同编程语言都可通过 http://localhost:18083/? wsdl 获取该 Webservice 的 WSDL 描述,继而生成相关的代理代码。下面是 C♯语言形式的代理代码。

```csharp
public class PTRService : System.Web.Services.WebService
{
    public struct PTRJobDetailInfo
    {
        //任务唯一 ID
        public string m_jobID;
        //任务提交用户名
        public string m_username;
        //Job 相关的计算节点组
        public string[] m_taskServers;

        //Job 相关子任务 ID
        public string[] m_tids;
        //Job 相关子任务进度
        public float[] m_tpercents;
        //任务调度者
        public string m_taskManager;

        //算法 ID
        public string m_algID;
        //算法输入参数
        public string m_inputParamsD;

        //任务优先级
        public int m_priority;
        //任务超时阈值
        public int m_timeOut;

        //任务开始时间
```

```csharp
    public string m_beginTime;
    //任务提交时间
    public string m_submitTime;
    //任务结束时间
    public string m_endTime;
}
public struct PTRJobRunInfo
{
    //任务唯一ID
    public string m_jobID;
    //执行进度
    public float m_percent;
    //任务状态,1表示正在运行;2表示排队等待;3表示任务完成;4表示任务
    //执行失败
    public int m_status;
    //任务提交用户
    public string m_username;
    //任务类型(保留)
    public string m_jobType;

    //任务信息(保留)
    public string m_jobInfo;
}
public struct PTRWorkInfo
{
    //工作进程唯一ID
    public string wuuid;
    //工作进程类型,1表示子任务执行者,2表示任务调度者
    public int workerType;
    //OpenRS算法唯一ID
    public string algID;

    //任务类型(保留)
    public string jobType;
    //任务唯一ID
    public string jobID;
    //任务状态,1表示正在运行;2表示排队等待;3表示任务完成;4表示任务执行
    //失败
```

```csharp
    public int jobStatus;
    //任务执行进度(0 到 1 之间)
    public float jobProcessInfo;

    //任务提交用户名
    public string username;
    //工作进程所在主机
    public string hostname;

    //工作进程消耗的 CPU
    public float cpuUsage;
    //工作进程消耗的内存
    public int memory;

    //是否繁忙
    public int isBusy;
};

public struct PTRMachineInfo
{
    //机器名
    public string hostname;
    //机器核数
    public int nProcessors;
    //机器内存总数
    public int memoryTotal;
    //可用内存数
    public int memoryFreeTotal;
    //磁盘总可用容量
    public int diskFreeTotal;
    //磁盘总容量
    public int diskTotal;
    //CPU 总利用率
    public float cpuUsage;
};

public struct PTRMachineDetailInfo
{
```

```
    public PTRMachineInfo minfo;
    public PTRWorkInfo []servInfos;
};

//改变任务优先级
int changePriority(string jobID, int priority);

//当用户发出任务取消请求后,第一步是任务管理与调度模块接收,然后将取消指
//令发送给任务调度者,由任务调度者中止当前正在执行的远程进程
int killJob(string jobID);

//任务运行时信息查询:任务运行时信息主要包括任务执行过程中的警告、错误等信
//息输出,任务状态
//(失败、运行中、挂起、等待),任务的子作业个数,任务进度,任务参数信息,子任务
//进度
//获取任务简略信息
int getRunInfo(string jobID,out PTRJobRunInfo info);
//获取任务详细信息
int getDetailInfo(string jobID,out PTRJobDetailInfo info);
//获取所有任务信息
int getAllJobRunInfos(out PTRJobRunInfo[] infos);

//机器状态信息主要包括机器名,机器 IP,机器核数,当前运行任务,硬盘容量,内存
//容量,CPU 繁忙状态等
//获取所有机器简略信息
int getAllMachineInfo(out PTRMachineInfo[] infos);
//获取某个机器详细信息
int getMachinDetailInfo(string machineName, out PTRMachineDetailInfo info)
}
```

8.3 PTR 并行编程接口模型

8.3.1 接口定义

PTR 并行编程接口按照并行粒度不同分为三类接口,分别为简单任务接口、分解与合并任务接口及层次子任务接口。算法编写者根据不同的算法并行特性,选择三类接口之一,按照接口抽象定义进行算法实现。下面将对每种接口分别进

行阐述。

1. 分解与合并接口(orsIParallelExe)

orsIParallelExe 覆盖的遥感处理更广，orsISimpleExe 可看成它的一个特例。该类的特点是任务分解后形成多个子任务，子任务合并时需要特殊处理。接口定义如下：

```
interface orsIParallelExe : public orsIExecute
{
public:
    //若 nTasks = 0, 则算法自己决定任务数,否则按照给定的任务数运行
    virtual orsArray<ref_ptr<orsIProperty>> getTasks(int nTasks = 0) = 0;
    //通过 Job 信息和 Task 信息进行计算得到输出信息,taskOutput 在外部分配内存
    //<map>
    virtual ref_ptr<orsIProperty> taskExecute(ref_ptr<orsIProperty> taskInput, or-
        sIProcessMsg * process) = 0;
    //进行子任务合并<reduce>
    virtual ref_ptr<orsIProperty> taskCombine(orsArray<ref_ptr<orsIProperty>>
        taskInputs, orsIProcessMsg * process) = 0;

    ORS_INTERFACE_DEF(orsIExecute, _T("parallel"));
};
```

分解与合并类算法的特点是处理过程可分为任务分解、子任务处理与子任务合并三个阶段，每个阶段按照算法特点不同而不同。很多遥感影像处理算法都属于分解与合并类，如下所述。

(1) 监督分类：监督分类算法在任务分解阶段主要包含模型训练与按照影像范围生成子任务；子任务执行阶段为使用训练模型逐像素分类，并形成各子任务对应的分类结果；子任务合并阶段为将分类结果进行合并形成最终整体分类图。主要包括最大似然、马氏距离、二值编码、支持向量机、神经网络、免疫分类等。

(2) 边缘检测：这里的边缘检测算法主要指最终结果为单像素精度线状地物的边缘检测算法，如 Canny、Edison 等。

(3) 几何纠正：按影像范围进行子任务并行处理，但子任务处理完毕后，需根据地理参考进行较为复杂的合并过程(插值、接边)，而不仅仅是文件合并，包括多项式纠正、RPC 纠正、各种严密成像方程纠正等。

(4) 图像融合：指使用两种影像进行像素级融合的算法。包括 Brovey、Multi-

plicative、HIS、PanSharp。

(5) 非全局约束的影像配准与匹配：指并不考虑匹配点检全局约束，仅采用匹配点的像素邻域进行相关系数或互信息熵的方法。包括各种波段配准、INSAR 处理中配准等。

(6) 其他：包括一些现有的并行算法，如并行 Mean Shift，即采用影像范围进行子任务处理，然后进行拼接；Snaphu 相位解缠对每个小影像块分别进行解缠得到各自的网络图，然后利用全局信息形成二级全局网络，进行全局解缠。

2. 层次子任务接口（orsIParallelExe_L）

层次子任务的特点是各个子任务时间具有树状层次关联关系，一个子任务的执行需前几个子任务完成后才能触发。分解与合并接口可以看成其只有一个层次的子任务的特例。

```
interface orsIParallelExe_L : public orsIExecute
{
public:
    virtual ors_int32 getLevels(ors_int32 * totalTasks, int nTasks = 0) = 0;

    //若 nTasks = 0, 则算法自己决定任务数,否则按照给定的任务数运行
    virtual orsArray〈ref_ptr〈orsIProperty〉〉 getTasks(int iLevel) = 0;
    //通过 Job 信息和 Task 信息进行计算得到输出信息,taskOutput 在外部分配内存
    〈map〉
    virtual ref_ptr〈orsIProperty〉 taskExecute(int iLevel, ref_ptr〈orsIProperty〉
        taskInput,orsIProcessMsg * process) = 0;
    //进行子任务合并〈reduce〉
    virtual ref_ptr〈orsIProperty〉 taskCombine(int iLevel, orsArray〈ref_ptr〈or-
        sIProperty〉〉 taskInputs,orsIProcessMsg * process) = 0;

    ORS_INTERFACE_DEF(orsIExecute, _T("parallel_l"));
};
```

该接口模型主要用于某些复杂算法。典型的例子是多分辨率影像匹配，每一级的匹配需要上一级匹配的结果作为初值。

8.3.2 数据交换与信息输出

目前，PTR 并行抽象接口模型暂不支持子任务间的通信。每个任务的输入与

输出均用 orsIProperty 表示。值得注意的是，算法中的 taskExecute、taskCombine、getTasks 等函数一般不在同一主机上执行，因此这些函数的输入不能为全局变量，仅为其函数输入参数。每个子任务的输入采用 orsIProperty 表示，getTasks 获得的子任务（orsIProperty）会由系统传递到 taskExecute 和 taskCombine。

子任务在执行与合并过程中可以进行信息输出，一方面可以方便调试；另一方面在错误发生时（如输入参数错误），可反馈给 PTR 并行处理框架，从而进行错误处理并反馈给网络用户。信息输出为 orsIProcessMsg 接口，接口形式如下：

```
typedef enum orsLogLEVEL{
    ORS_LOG_DEBUG,//调试输出
    ORS_LOG_INFO,//一般信息输出
    ORS_LOG_WARNING,//警告信息输出
    ORS_LOG_ERROR,//错误信息输出
    ORS_LOG_FATAL//致命错误信息输出
}orsLogLEVEL;

interface orsIProcessMsg : public orsIObject
{
public:
    //进度,在0.0到1.0之间, 返回零表示取消任务
    virtual bool process(double status) = 0;
    //消息打印
    virtual int logPrint(orsLogLEVEL loglevel, const orsChar * fmt, …) = 0;

    //接口名称
    ORS_INTERFACE_DEF(orsIObject, _T("process"))
};
```

其中，process 方法主要输出进度信息，logPrint 方法输出日志消息，当日志消息类型为 ORS_LOG_ERROR 或 ORS_LOG_FATAL 时，系统会中断后续算法过程，反馈给用户相关错误提示。在算法中，常用日志接口使用模板式代码如下：

```
//传入的进度对象
orsIProcessMsg * process ;

//输出错误信息
int erroroid = 34;
```

```
const char * reseaon = "input file is empty";
process->logPrint(ORS_LOG_ERROR, "orsPLExample: %d(%s)", errorid, reseaon);
```

```
//输出进度
process->process(0.5f);
```

8.3.3 并行算法插件实例

下面将对每种并行模型分别给出简单算法插件实例。

1. 简单任务算法实例

该算法用于影像 DCT 变换,下面列出部分代码,完整代码参见"\OpenRs\desktop\src\plugins\orsParallelExample\orsSE_Dct.cpp"代码。

```cpp
class orsSE_Dct:public orsIExeHelper<orsISE_ImageOrthoTrans> public orsObjectBase
{
private:
    orsStringm_inputImageFileName;
    orsStringm_outputImageFileName;

public:
    orsSE_Dct(bool bForRegister)
    :orsIExeHelper<orsISE_ImageOrthoTrans>(bForRegister)
    {
        if (!bForRegister) {
            m_inputImageFileName = "Input Image File";
            m_outputImageFileName = "Output Image File";

            m_inputFileNames->addAttr("inputImageFile", m_inputImageFileName);
            m_outputFileNames->addAttr("outputImageFile", m_outputImageFileName);

        }
    }

    virtual bool setArguments(orsIProperty * inputFileNames, orsIProperty * parame-
        terArgs, orsIProperty * outputFileNames)
    {
```

```cpp
        inputFileNames->getAttr("inputImageFile", m_inputImageFileName);
        outputFileNames->getAttr("outputImageFile", m_outputImageFileName);
        return true;
    }

    virtual ref_ptr<orsIProperty> execute(orsIProcessMsg *process)
    {
        //根据 m_inputImageFileName 与 m_outputImageFileName 进行处理
        //代码略
    }

    ORS_OBJECT_IMP3(orsSE_Dct, orsISE_ImageOrthoTrans, orsISimpleExe, orsIExecute, _T("Dct"), _T("Simple Dct"))
};
#define ORS_SE_ORTHOTRANS_DCT _T("ors.execute.simple.Dct")
```

2. 分解与合并处理算法实例

该算法用于并行中值滤波，下面列出部分代码，完整代码参见"\OpenRs\desktop\src\plugins\orsParallelExample\orsPE_MedianFilter.cpp"。

```cpp
class orsPE_MedianFilter : public orsIExeHelper<orsIPE_ImageFilter>, public orsObjectBase
{
private:
    orsString m_inputImageFileName;
    orsString m_outputImageFileName;

public:
    orsPE_MedianFilter::orsPE_MedianFilter(bool bForRegister)
    :orsIExeHelper<orsIPE_ImageFilter>(bForRegister)
    {
        if(!bForRegister) {
            m_inputImageFileName = "Input Image File";
            m_outputImageFileName = "$(inputImageFile)_mf.tif";

            m_inputFileNames->addAttr("inputImageFile", m_inputImageFileName);
            m_outputFileNames->addAttr("outputImageFile", m_outputImageFileName);
```

```cpp
        }
    }
    //参数设置
    virtual bool setArguments(orsIProperty *inputFileNames, orsIProperty *parame-
        terArgs, orsIProperty *outputFileNames)
    {
        inputFileNames->getAttr("inputImageFile", m_inputImageFileName);
        outputFileNames->getAttr("outputImageFile", m_outputImageFileName);

        return true;
    }
    //获取并行子任务
    virtual orsArray<ref_ptr<orsIProperty> > getTasks(int nTasks)
    {
        orsArray<ref_ptr<orsIProperty> > tasks;

        GDALDatasetH hDataset = GDALOpen(m_inputImageFileName, GA_ReadOnly);
        long imgWid = GDALGetRasterXSize(hDataset);
        long imgHei = GDALGetRasterYSize(hDataset);
        GDALClose(hDataset);

        int fromRow = 0;
        int subRows = (imgHei + nTasks - 1) / nTasks;
        for(int i=0; i< nTasks; i++)
        {
            ref_ptr<orsIProperty> task = getPlatform()->createProperty();
            if(i == nTasks-1)
                subRows = imgHei - fromRow;
            //子任务起始、终止行
            task->addAttr("task:fromRow", (ors_int32)fromRow);
            task->addAttr("task:toRow", (ors_int32)(fromRow + subRows));

            tasks.push_back(task);
            fromRow += subRows;
        }

        return tasks;
    }
```

```cpp
//子任务执行
virtual ref_ptr<orsIProperty> taskExecute(ref_ptr<orsIProperty> taskInput,orsIProcessMsg* msg)
{
    msg->process(0.0);
    msg->logPrint(ORS_LOG_INFO,"task begin");
    ors_int32 fromRow;
    ors_int32 toRow;
    //取子任务起始、终止行
    taskInput->getAttr("task:fromRow", fromRow);
    taskInput->getAttr("task:toRow", toRow);
    //子任务结果输出文件
    orsString taskfile_out;
    taskfile_out=GetTaskTempFileName(m_outputImageFileName);
    int pos=taskfile_out.reverseFind('.');
    ref_ptr<orsIProperty> task = getPlatform()->createProperty();

    if(pos > -1) {
        taskfile_out = taskfile_out.left(pos);
        //形成子任务输出文件名
        char buf[256];
        sprintf(buf, "%s_%d.tif", taskfile_out.c_str(), fromRow);
        taskfile_out = buf;

        //子任务起始、终止行、输出文件名
        task->addAttr("task:fromRow",fromRow);
        task->addAttr("task:toRow",toRow);
        task->addAttr("task:file_out",taskfile_out);

        msg->logPrint(ORS_LOG_INFO,"task file out : %s",taskfile_out.c_str());

        //调用 orsSE_MedianFilter 执行子任务,这样做是为了重用 orsSE_median
        //Filter
        {
            ref_ptr<orsIProperty> subJobPara = getPlatform()->createPro-
```

```cpp
            perty();
        //准备调用参数
        subJobPara->addAttr("inputImageFile", m_inputImageFileName.c_str
            ());
        subJobPara->addAttr("outputImageFile", taskfile_out.c_str());
        subJobPara->addAttr(ORS_RROI_BEGINROW, fromRow);
        subJobPara->addAttr(ORS_RROI_NUMOFROWS, toRow-fromRow);
        ref_ptr<orsISimpleExe> filterExe = ORS_CREATE_OBJECT(orsISimpleExe,
            ORS_SE_IMAGEFILTER_MEDIAN);

        if(NULL != filterExe.get()) {
            filterExe->initFromProperty(subJobPara.get());
            filterExe->execute(msg);
        }
    }
    else {
        getPlatform()->logPrint(ORS_LOG_ERROR, "Invalid output image file name");
    }

    return task;
}
    //任务合并
    virtual ref_ptr<orsIProperty> taskCombine(orsArray<ref_ptr<orsIProperty>
        > taskInputs, orsIProcessMsg * msg)
    {
        ref_ptr<orsIProperty> jobOut = getPlatform()->createProperty();
        jobOut->addAttr("job:tasknum", (ors_int32)taskInputs.size());
        jobOut->addAttr("filtered file", m_outputImageFileName);

        msg->process(0);
        msg->logPrint(ORS_LOG_INFO, "开始合并影像");
        //代码略

        return jobOut;
    }
```

```
        ORS_OBJECT_IMP3(orsPE_MedianFilter, orsIPE_ImageFilter, orsIParallelExe,
            orsIExecute, _T("medianFiler"), _T("Distributed Median Filter"))
};
#define ORS_PE_IMAGEFILTER_MEDIAN _T("ors.execute.parallel.imageFilter.median")
```

3. 层次子任务算法实例

该算法为层次子任务算法例子,下面列出部分代码,完整代码参见"\OpenRs\desktop\src\plugins\orsParallelExample\orsPL_Test.cpp"。

```
class orsPLExample : public orsIExeHelper<orsIParallelExe_L>, public orsObjectBase
{
private:
    orsStringm_inputImageFileName;
    orsStringm_outputImageFileName;

public:
    orsPLExample::orsPLExample(bool bForRegister)
    :orsIExeHelper<orsIParallelExe_L>(bForRegister)
    {
        if(!bForRegister) {
            m_inputImageFileName = "Input Image File";
            m_outputImageFileName = "$(inputImageFile)_mf.tif";

            m_inputFileNames->addAttr("inputImageFile", m_inputImageFileName);
            m_outputFileNames->addAttr("outputImageFile", m_outputImageFileName);
        }
    }
    //参数设置
    bool orsPLExample::setArguments(orsIProperty * inputFileNames, orsIProperty *
        parameterArgs, orsIProperty * outputFileNames)
    {
        inputFileNames->getAttr("inputImageFile", m_inputImageFileName);
        outputFileNames->getAttr("outputImageFile", m_outputImageFileName);

        return true;
    }
```

```cpp
//取层数
ors_int32 orsPLExample::getLevels(ors_int32 * totalTasks, int nTasks)
{
    //每层 4 个 map 任务，一共 5 层
    ors_int32 nlevel = 3;
    * totalTasks = 3 + 4 + 2;

    return nlevel;
}
//取第 iLevel 层任务
orsArray<ref_ptr<orsIProperty> > orsPLExample::getTasks(int iLevel)
{
    orsArray<ref_ptr<orsIProperty> > ps;

    int nTasks;
    switch(iLevel)
    {
    case 0:
        nTasks = 3;
        break;
    case 1:
        nTasks = 4;
        break;
    case 2:
        nTasks = 2;
        break;
    }

    for (int i=0; i<nTasks; i++)
    {
        ref_ptr<orsIProperty> mp = getPlatform()->createProperty();
        mp->addAttr("Level",(ors_int32)iLevel);
        mp->addAttr("ID",(ors_int32)i);
        ps.push_back(mp);
    }

    return ps;
}
```

```cpp
//任务执行
ref_ptr<orsIProperty> orsPLExample::taskExecute(int iLevel, ref_ptr<orsIProp-
    erty> taskInput,orsIProcessMsg *process)
{
    ors_int32 mLevel = 0;
    ors_int32 id = 0;
    taskInput->getAttr("Level", mLevel);
    taskInput->getAttr("ID", id);

    if(mLevel != iLevel)
    {
        getPlatform()->logPrint(ORS_LOG_ERROR,"erorr12");
        return NULL;
    }
    else {
        //模拟执行
        for (int i=0;i<200;i++)
        {
            ::Sleep(10);
            if(process != NULL)
                process->process((i+1.0)/200.f);
        }
    }
    ref_ptr<orsIProperty> mp = getPlatform()->createProperty();
    mp->addAttr("ID",id);
    return mp;
}
//任务合并
ref_ptr<orsIProperty> orsPLExample::taskCombine(int iLevel, orsArray<ref_ptr
    <orsIProperty> > taskInputs,orsIProcessMsg *process)
{
    int nTasks;
    switch(iLevel)
    {
    case 0:
        nTasks = 3;
        break;
```

```
            case 1:
                nTasks = 4;
                break;
            case 2:
                nTasks = 2;
                break;
        }

        if(taskInputs.size() != nTasks){
            getPlatform()->logPrint(ORS_LOG_ERROR, "orsPLExample: erorr2222");
            return NULL;
        }

        for (int i=0;i<200;i++)
        {
            ::Sleep(10);
            if(process != NULL)
                process->process((i+1.0)/200.f);
        }

        ref_ptr<orsIProperty> mp = getPlatform()->createProperty();

        return mp;
    }

    ORS_OBJECT_IMP2(orsPLExample, orsIParallelExe_L, orsIExecute, _T("example"), _
        T("Parallel Level Example"))
};
```

8.4 基于ROI属性的自动并行机制

8.4.1 遥感数据并行处理的特点

通过对遥感通用处理算法的总结与分析,根据对内存中数据处理方式的不同,遥感数据处理算法可以归纳为五大类。

(1) 单点运算。

包括灰度反转、指数变换、对数变换、线性变换、阈值变换、窗口变换、直方图均衡等。这类算法的特点是,像素点灰度变换的规则一旦定义,运算后新像素的值可以唯一确定。

(2) 邻域运算。

包括边缘检测、图像平滑、中值滤波、图像腐蚀、图像膨胀等。这类算法的特点是输出的像素值由周围像素决定。

(3) 正交变换(二维可分)。

包括 FFT、小波变换、离散余弦变换、沃尔什变换等。这类算法的特点是需要先对行进行变换,然后再对列进行变换。

(4) 多波段运算。

线性运算:包括 KL 变换、RGB/HIS 彩色变换、缨帽变换等。这类算法的特点是输出的像素是由该点在不同波段上灰度值运算而获取的。

非线性运算:指数类运算等。

(5) 几何变换。

包括图像平移、镜像变换、转置变换、缩放变换和旋转变换。这类算法的特点是给出一个分块的内存区域,返回一个新的内存分块区域。

其中,大部分算法都可以进行局部计算。如图 8-9 所示,对于这种仅依赖局部信息的栅格数据处理,其并行模式可以分为三类:

① 按行列分块;
② 按行分块,列保持为影像原始宽度;
③ 按列分块,行保持为影像原始高度。

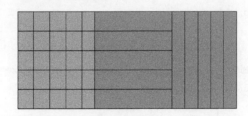

图 8-9　遥感影像处理的三种分块模式

如果把这种栅格数据的处理写成并行模式,那么会发现大部分并行化的代码都是相同的。如果每个算法都重新写一遍,那么会让人觉得很枯燥乏味,也增加了调试的工作量。如果能够设计一种针对这种栅格数据局部处理的并行模式,那么可以大大减少编程的工作量,让算法开发人员专注于算法本身的开发,提高编程的效率。

简单任务接口(orsISimpleExe)主要针对任务级并行,没有子任务分解合并过程,相对简单。接口定义如下:

```
//简单任务接口
interface orsISimpleExe: public orsIExecute
{
public:
    //计算得到输出参数
    virtual ref_ptr<orsIProperty> execute(orsIProcessMsg * process) = 0;
    ORS_INTERFACE_DEF(orsIExecute, _T("simple"));
};
```

简单任务可执行对象可通过 ROI 属性和 ORS_RASTER_EXE_OBJECT 代理对象转换为并行处理对象(orsIParallelExe)。ORS_RASTER_EXE_OBJECT 代理对象通过设置简单任务可执行对象的 ROI 属性，将影像按照影像范围进行子任务分解，各子影像块完成后，再将其合并为最终影像，实现子任务分解与合并自动化。因此，不能认为简单任务接口不能并行。事实上，OpenRS 大部分算法为支持 ROI 属性的 orsISimpleExe 对象，如图 8-10 所示。

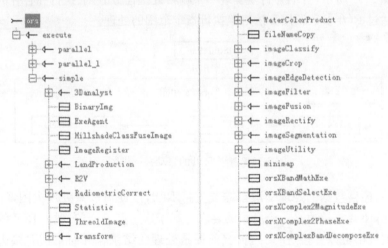

图 8-10　orsISimpleExe 算法树

有一类遥感处理算法的特点是以影像块为单位进行任务分解，块内每个像素处理仅涉及邻近小范围的若干像素，而任务合并的主要过程为中间影像文件的拼接，无复杂的合并处理过程，可称为"非整体局部处理"类算法。对于这类算法均可按照 orsISimpleExe＋ROI 的方式实现并行。具体而言，主要包括如下算法。

（1）邻域类图像处理：对图像进行逐像素处理，处理方式为以待处理像素为中心的邻近小范围内像素值为输入进行计算，如 HIS 变换、阈值计算、形态学计算、插值计算、模板处理。

(2) 滤波算法：指各种仅需使用邻域信息的滤波算法。包括中值滤波、均值滤波、Lee 滤波、增强 Lee 滤波、Frost 滤波、增强 Frost 滤波、GammaMap 滤波、PPB 滤波、三边滤波、干涉图滤波等。

(3) 高光谱定量遥感指数计算：主要包括仅使用单像素多光谱信息的各种定量遥感指数。包括 NDVI、EVI、WI、PRI、CARI、LAI、SIPI、CARI、APAR、PRI 等。

(4) 像素级特征计算：主要指采用模板方式进行各种像素级特征提取。包括均值方差计算、PSI 特征计算、波段算数运算（band math）等。

(5) 其他：相位转高程、干涉图生成、DEM 晕渲图计算。

8.4.2 简单任务与并行细分任务的统一

基于大部分栅格数据处理可局部化这一特点，可以设计一个自动并行器来完成栅格处理任务的分割与结果合并。实际上，如果需要进行自动并行化的可执行对象支持 ROI 操作，则可以很容易地编写公用的自动并行器来分块调用算法对象。如图 8-11 所示，自动并行器实现一个栅格处理任务分割、执行和合并的框架，执行时通过 ROI 参数调用算法对象执行指定范围的处理。

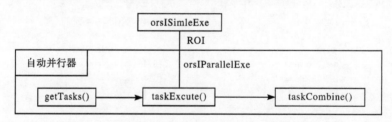

图 8-11　基于 ROI 的自动并行机制

另外，栅格数据的处理一般都需要支持 ROI，以便可以只处理大图像的一部分。因此，ROI 支持也是一个必须的要求。基于这一点，只要实现了 ROI 的算法对象原则上都可以直接进行并行化（前提是处理的结果合并时不需要接边）。从这种意义上说，ROI 属性实现了栅格数据简单处理对象和并行处理对象的一种统一。

在 OpenRS 中，栅格处理的可执行对象定义了栅格 ROI 属性 ORS_RASTER_ROI。栅格 ROI 属性包含起始行号、列号、行数、列数四个子属性。

```
//影像上的ROI
#define ORS_RASTER_ROI_T("ROI in Pixels")


//起始行
```

```
#define ORS_RROI_BEGINROW_T("BeginRow")
//行数
#define ORS_RROI_NUMOFROWS_T("NumOfRows")

//起始列
#define ORS_RROI_BEGINCOL_T("BeginCol")
//列数
#define ORS_RROI_NUMOFCOLS_T("NumOfCols")
```

8.4.3 基于 ROI 的自动并行化属性定义

为了支持自动并行化的扩展,必须为栅格对象提供指定自动并行对象的机制。为此定义了自动并行化的属性关键字 ORS_PARALLELIZE_OBJECT。

```
//简单对象对应的并行对象,或自动并行对象,关键字
#define ORS_PARALLELIZE_OBJECT_T("ParallelizeObject")
```

在栅格处理对象的属性列表中增加 ORS_PARALLELIZE_OBJECT 即可指定用于并行化的对象。OpenRS 中目前已经实现了一个面向像素坐标的自动并行化对象 ORS_PE_RASTER_PROC。

```
//栅格类处理自动并行器的缺省实现
#define ORS_PE_RASTER_PROC_T("ors.execute.parallel.rasterProc")
```

另外,为了支持栅格数据并行处理的三种模式,定义属性关键字 ORS_RASTER_EXE_PARALLEL_MODE 和三种并行模式的属性值。

```
//自动并行的模式,关键字
#define ORS_RASTER_EXE_PARALLEL_MODE_T("RasterExeParallelMode")

//自动并行的模式
#define ORS_RASTER_PE_byROW_T("RasterExeParallelByRow")
#define ORS_RASTER_PE_byCOL_T("RasterExeParallelByCol")
#define ORS_RASTER_PE_byTILE_T("RasterExeParallelByTILE")
```

一个典型的栅格处理可执行对象,只需要在构造函数中增加如下属性,就可以被框架程序感知并自动支持并行化(具体实例参看\OpenRS\desktop\src\plugins\orsParallelExample 目录下的 orsSE_MedianFilter.cpp)。

```
//设置自动并行对象
m_jobArguments->addAttr(ORS_PARALLELIZE_OBJECT, ORS_PE_RASTER_PROC);

//设置任务分割模式
m_jobArguments->addAttr(ORS_RASTER_EXE_PARALLEL_MODE, ORS_RASTER_PE_byTILE);
```

栅格ROI属性可执行对象在执行时,配置界面会出现如图8-12所示的ROI属性组和进程数选项。当进程数大于1时,框架程序将用自动并行化对象代替原本对象执行,并把原对象ID通过ORS_RASTER_EXE_OBJECT属性传递给自动并行化对象。

```
//要被自动执行的栅格处理对象,关键字
#define ORS_RASTER_EXE_OBJECT_T("RasterExeObject")
```

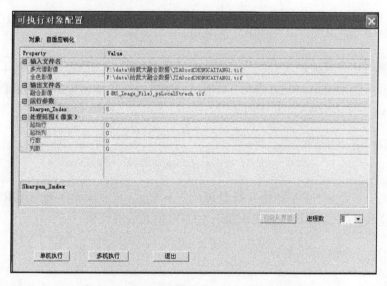

图8-12 基于ROI属性的自动并行配置界面

第9章 网络服务包装与嵌入应用

9.1 一键式网络服务包装

9.1.1 基本思想

微软 IIS 服务器直接支持 C♯ 源代码,通过[WebMethod]特性可以直接把 C♯ 的函数发布为标准的网络服务。因此,如果把 OpenRS 的可执行对象调用包装为一个具有[WebMethod]特性的 C♯ 函数,那么就可以直接把 OpenRS 的处理功能发布为标准网络服务。

如图 9-1 所示,可以利用 C♯ 程序 service.cs 把 OpenRS 的可执行对象 exeObject 自动包装为 asp.net 的 WebMethod 函数,然后通过 IIS server 发布为标准的网络服务。在调用时,IIS Server 把外部网络调用转换为 service.cs 中 WebMethod 函数的调用,进而提交给 OpenRS 的分布式处理任务管理服务器 OpenRS_Cloud Server。

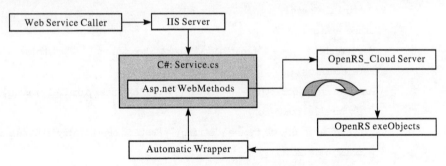

图 9-1 遥感数据处理模块的插件自动服务包装机制

9.1.2 具体实现

在实现时,OpenRS 可执行对象的调用将被转换为 C♯ 的 WebMethod 函数。以可执行对象 ors.execute.simple.BinaryImg 为例,对象 ID 将被转换为 C♯ 的函数名 OpenRS_BinaryImg,即 "ors.execute.simple." 替换为"OpenRS_";输入、输出文件名和阈值参数被转换为 OpenRS_BinaryImg 的调用参数(string InImageFileName, string outImageFileName, double threshold)。

服务调用时，OpenRS_BinaryImg 被调用。OpenRS_BinaryImg 所做的工作实质上是把调用参数(string InImageFileName，string outImageFileName，double threshold)自动打包成 XML 字符串的格式，然后提交到 OpenRS 的分布式处理任务管理器。代码示例如下：

```csharp
//C#网络服务函数特性声明
[WebMethod(Description = "return jobid.")]
public string OpenRS_BinaryImg(string OrderID, string InImageFileName, string outImageFileName, double threshold)
{
    string strXML =
        "〈OpenRS_Cloud〉
        〈OrderID type=\"string\"〉"+OrderID + "〈/OrderID〉
        〈InputFileNames〉
            〈InImageFileName type=\"string\"〉" + InImageFileName +"〈/InImageFileName〉
        〈/InputFileNames〉
        〈OutputFileNames〉
            〈OutImageFileName type=\"string\"〉"+OutImageFileName+"〈/OutImageFileName〉
        〈/OutputFileNames〉
        〈ParameterArgs〉
            〈Threshold type=\"float64\"〉"+para_Threshold+"〈/Threshold〉
        〈/ParameterArgs〉
        〈ParallelizeObject type=\"string\"〉ors.execute.parallel.rasterProc
        〈/ParallelizeObject〉
        〈RasterExeParallelMode type=\"string\"〉RasterExeParallelByTILE〈/RasterExeParallelMode〉
        〈ROI_in_Pixels〉
            〈BeginCol type=\"int32\"〉0〈/BeginCol〉
            〈BeginRow type=\"int32\"〉0〈/BeginRow〉
            〈NumOfCols type=\"int32\"〉0〈/NumOfCols〉
            〈NumOfRows type=\"int32\"〉0〈/NumOfRows〉
        〈/ROI_in_Pixels〉
        〈RasterExeObject type=\"string\"〉ors.execute.simple.BinaryImg〈/RasterExeObject〉
        〈Num_Of_Processes type=\"int32\"〉4〈/Num_Of_Processes〉
```

</OpenRS_Cloud>";

//PTR 服务提交对象
JobService.PTRService sr = new JobService.PTRService();

sr.Url = "http://localhost:18083/?wsdl";
int ret = sr.submitJob("guowei", " ", " ", "ors.execute.parallel.rasterProc",
    strXML, 1, 30000, out jobID);

string jobID = "";

return jobID;
}

9.1.3 一键自动包装

在可执行对象执行器模块 orsExeRunner 中,设置了"WriteWSDL"命令实现一键式可执行对象包装为网络服务的功能。

"WriteWSDL"会自动检查 etc/webService/目录下是否存在文件 OpenRS_WebService.txt,若存在则会从中提取需要包装的可执行对象,若 OpenRS_WebService.txt 文件不存在,则默认为所有的可执行对象都需要包装(图 9-2)。

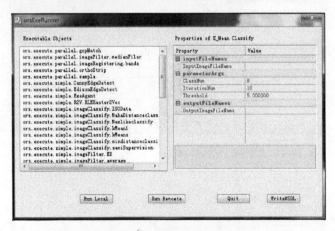

图 9-2 遥感数据处理模块的插件自动服务包装

OpenRS_WebService.txt 中列出了需要自动包装的对象 ID,如:

ors.execute.simple.3Danalyst.SurfaceAnalyst.hillshade
ors.execute.simple.BinaryImg

ors.execute.simple.HillshadeClassFuseImage

ors.execute.simple.Statistic

ors.execute.simple.ThreoldImage

ors.execute.simple.imageClassify.supervised.SVM

ors.execute.simple.imageFilter.median

9.1.4 实现效果

服务发布到 IIS 网站后，就可以在浏览器中查询到自动包装的所有可执行对象（图 9-3），可按照标准网络服务调用执行（图 9-4）。

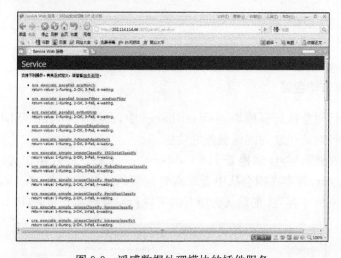

图 9-3　遥感数据处理模块的插件服务

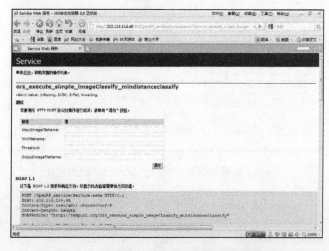

图 9-4　遥感数据处理模块的插件服务调用

9.2 嵌入应用技术

OpenRS处理服务是基于OpenRS桌面处理、分布式并行处理、网络处理服务一体化架构开发的遥感数据网络处理服务。根据目前开发和部署的情况，OpenRS处理服务与门户网站和编目系统的接口关系如图9-5所示。OpenRS接收用户在门户网站提交的产品订单后按照OpenRS内部网络服务的方式转发到PTR并行处理管理与调度器（ManegerServed），然后PTR系统调用节点处理（WorkServed）进行数据处理生产，生产的数据输出到Output目录，最后产品通知消息直接通过网络服务提交到编目系统，也可以空文件的形式输出到notice目录。

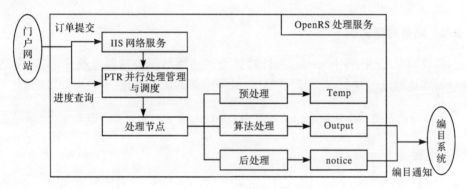

图9-5　OpenRS服务处理流程

OpenRS数据处理节点（WorkServerd）接收到服务请求后，按照下述流程对数据进行处理：

（1）数据预处理。从XML请求中提取参数，包括订单号、输入数据、输出数据、算法参数及算法对象等相关信息。根据场景配置信息，将输入数据和输出数据的地址进行映射（URL地址到本地地址）。然后，判断输入数据是否需要解压缩，并将预处理后的结果存放在配置信息指定的目录中。

（2）数据处理。根据提取出来的服务请求参数，调用相应的算法对象，对经过预处理的数据进行处理，并将产品数据保存在场景配置信息指定的目录。

（3）数据后处理。根据场景配置信息，对生成的产品数据进行压缩处理，生成编目打包文件，并通过网络服务通知编目系统。

9.2.1　场景对象接口

为了适应不同场景的预处理和后处理，定义了并行处理场景对象orsIParallelScene 接口。这样在具体执行和算法有关的处理前后,可以根据需要执行预处理和后处理操作。

```
interface orsIParallelScene: public orsIObject
{
    virtual bool PreProcess(orsIProperty * inProperty, orsIProperty * outProperty,
        orsIProcessMsg * processMsg) = 0;
    virtual bool PostProcess (orsIProperty * inProperty, orsIProcessMsg * processMsg) = 0;
public:
    ORS_INTERFACE_DEF(orsIObject, _T("parallelScene"));
};
```

9.2.2 场景对象实例

场景对象 orsParalleScene_A 针对典型的遥感数据处理服务场景,实现网络地址到本地地址的映射、数据包的解压、处理结果的压缩打包、编目通知等功能。

```
class orsParalleScene_A: public orsIParallelScene, public orsObjectBase
{
private:
    //配置文件
    orsSceneConfig m_sceneConfig;

private:
    //解压缩,可能包含多个文件
    bool UnZip(orsIProperty * zipFile,  orsIProcessMsg * processMsg);
    //文件压缩,可能包含多个文件
    orsString ZipFiles(orsIProperty * outFiles,  orsIProcessMsg * processMsg);

public:
    orsParalleScene_A(bool bForRegister);
    ~orsParalleScene_A();

    //预处理
    bool PreProcess(orsIProperty * inProperty, orsIProperty * outProperty, orsIProcessMsg * processMsg)
    {
        m_sceneConfig.Init();
```

//http 文件路径转换为本地路径

outProperty->copy(inProperty);

ref_ptr<orsIProperty> inputFileNames;
ref_ptr<orsIProperty> parameterArgs;
ref_ptr<orsIProperty> outputFileNames;
ref_ptr<orsIProperty> roiArgs;

outProperty->getAttr(INPUT_FILE_NAMES, inputFileNames);
outProperty->getAttr(PARAMETER_ARGS, parameterArgs);
outProperty->getAttr(OUTPUT_FILE_NAMES, outputFileNames);

//把 URL 地址转换为本地路径
m_sceneConfig.WebToLocal(inputFileNames.get());

//ZIP 文件解压
UnZip(inputFileNames.get(), processMsg);

//替换输出文件名
getUtilityService()->ReplaceMacroString(outputFileNames.get(), input-
    FileNames.get(), parameterArgs.get());

//把输出文件名改成特定的目录
m_sceneConfig.ChangeOutputDir(outputFileNames.get());

processMsg->process(1.0);

return true;
}

//后处理
bool PostProcess(orsIProperty * inProperty, orsIProcessMsg * processMsg)
{
    m_sceneConfig.Init();
```
```

```cpp
ref_ptr<orsIProperty> outProperty;

outProperty = getPlatform()->createProperty();

outProperty->copy(inProperty);

ref_ptr<orsIProperty> inputFileNames;
ref_ptr<orsIProperty> parameterArgs;
ref_ptr<orsIProperty> outputFileNames;
ref_ptr<orsIProperty> roiArgs;

outProperty->getAttr(INPUT_FILE_NAMES, inputFileNames);
outProperty->getAttr(PARAMETER_ARGS, parameterArgs);
outProperty->getAttr(OUTPUT_FILE_NAMES, outputFileNames);

orsString orderID;
orderID = _T("Unknown");
outProperty->getAttr(_T("OrderID"), orderID);

orsString outFileName;
outFileName = ZipFiles(outputFileNames.get(), processMsg);

bool bByNotifyFile = (m_sceneConfig.GetNotifyURL().length() < 4);
//写通知文件?
if(bByNotifyFile){
    orsString notifyFileName;
    notifyFileName = m_sceneConfig.GetNotifyPath() + _T("/");
    notifyFileName = notifyFileName + orsString::getPureFileName(out-
        FileName.c_str());

    notifyFileName += _T(".");
    notifyFileName += orderID + _T(".ok");

    FILE *fp = fopen(notifyFileName, "wt");

    if(NULL != fp)
        fclose(fp);
```

```
        }
        else{//直接使用 Web Service 通知
            ...
        }

        processMsg->process(1.0);
        return true;
    }

    private:
        ors_string m_pPreNetFilePath,m_pPreLocalFilePath;
        ors_string m_pPostNetFilePath,m_pPostLocalFilePath;

        ORS_OBJECT_IMP1(orsParalleScene_A,orsIParallelScene,"A","ParalleScene_A")
    };
```

其中,orsSceneConfig 为场景配置对象。orsSceneConfig 读取 Web Service 的场景配置文件,对处理节点进行配置。配置文件中包括 URL 地址到本地地址的映射、临时文件、产品文件及通知文件在本地的存储路径。场景配置文件一般存储在 OpenRS 目录 etc\WebService 下。典型的配置文件包括如下内容:

```
URLHead: http://localhost
LocalMapPath:e:\data

TempPath:e:\data\WebWork\temp
OutputPath:e:\data\WebWork\output
NotifyPath:e:\data\WebWork\notices

NotifyURL: http://192.168.10.16:8087/802Server/services/Order
```

其中,URLHead 为输入文件的 Web 地址前缀,LocalMapPath 为对应的本地目录;TempPath 为临时文件目录,OutputPath 为输出文件目录,NotifyPath 为处理结束通知文件目录。NotifyURL 为处理结束通知服务地址。

第 10 章 插件开发实践

10.1 插件开发的粒度

根据 OpenRS 的开放性设计,目前 OpenRS 支持可执行程序模块、可执行对象、处理算法和界面扩展四种粒度的插件开发。

(1) 可执行程序模块级。

以可执行程序快捷方式的形式集成到 OpenRS 主控界面。从本质上来说,这并不是真正的插件,而只是根据目录下的快捷方式,自动生成主界面菜单。参见 6.1 节 OpenRS 主控模块。

(2) 可执行对象级。

可执行对象是包含完整文件输入、输出的对象,可以以插件对象的形式集成到可执行对象执行器、orsViewer 动态菜单和工作流节点。参见 6.2 节对象执行器。

(3) 算法级。

以内存为主要输入、输出的算法或影像链节点对象,可以被其他对象调用。

(4) 界面扩展点。

以界面扩展的形式集成到 orsViewer 等定义特点扩展点的模块中,支持菜单、工具条、控件窗口的动态插入。参见 5.2 节。

10.2 粗粒度插件——可执行对象

可执行对象是继承 orsIExecute 接口的对象,相当于带有命令行参数的可执行程序,是 OpenRS 最粗粒度的对象。其中,根据并行特性又分为 orsISimpleExe、orsIParallelExe、orsIParallelExe_L 三种。orsISimpleExe 是不进行任务分裂的简单可执行对象接口;orsIParallelExe 是需要进行并行处理的对象接口;而 orsIParallelExe_L 是多层次并行处理对象接口。下面仅以 orsISimpleExe 和 orsIParallelExe 进行说明,关于 orsIParallelExe_L 的说明请参看第 8 章。

10.2.1 orsISimpleExe 与 orsIParallelExe

orsISimpleExe 是指一般不再细分的可执行功能,算法的输入和输出为具有

XML 类似层次接口的属性。在计算过程中,算法可以输出进度和日志信息,这些信息将直接反馈给 Web 用户。

orsIParallelExe 是指那些可以进行子任务分解和合并的并行算法,这些算法分解的子任务将会并行运行在由多任务并行作业系统所分配的节点机上,每个子任务依然可以输出进度和日志信息,orsISimpleExe 可以看成 orsIParallelExe 的特例。

在 OpenRS 系统中,算法参数的输入和输出采用 XML 字符串描述。因为采用 XML 参数描述,所以可以让 OpenRS-Cloud 系统自动地生成参数界面。目前 OpenRS 插件系统内置支持整数、浮点、字符串、文件名、数组和矩阵几种类型,并且可方便地扩展到空间点、线、面和用户自定义类型的支持,Web 入口系统可根据参数类型生成相应的界面元素。

从上述描述可以看出,算法实现者并不需要关心网络通信与资源分布情况,只需要实现算法逻辑即可。实际上实现 orsISimpleExe 或者 orsIParallelExe 的算法不仅可在 OpenRS-Cloud 中运行,也可以在桌面版的 OpenRS 系统中运行,这正是采用纯接口的方式而带来的网络化与桌面化运行的无缝集成(图 10-1)。

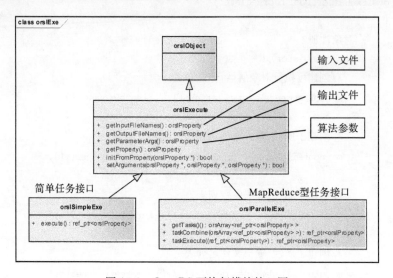

图 10-1　OpenRS 可执行模块接口图

10.2.2　对象命名规则建议

在 OpenRS 中可以根据接口名查询到实现该接口的对象,并由此构成对象 ID 列表,用于自动生成菜单或进行处理。为了把处理功能进行有序分类,必须对 orsISimpleExe 或 orsIParallelExe 进行细化。例如:

```
orsISimpleExe: orsIFilter_SE, orsIClassify_SE, …
orsIParallelExe: orsIFilter_PE, orsIClassify_PE, …
```

10.2.3 可执行对象帮助模板

为了简化开发，作者编写了若干可执行对象帮助模板。这些模板定义了标准的输入、输出、处理参数和 ROI 等属性变量，既简化了开发工作量，也统一了可执行对象的实现风格。

1. 普通可执行对象帮助模板

orsIExeHelper 模板用于实现通用的处理。

```cpp
class orsIExeHelper: public orsIExeInteface
{
protected:
    orsIExeHelper(bool bForRegister)
    {
        //对象注册?
        if(!bForRegister) {
            //创建可执行对象属性:作业参数
            m_jobArguments = getPlatform()->createProperty();

            //创建子属性:输入文件名列表
            m_inputFileNames = getPlatform()->createProperty();

            //创建子属性:算法参数列表
            m_parameterArgs = getPlatform()->createProperty();

            //创建子属性:输出文件名列表
            m_outputFileNames = getPlatform()->createProperty();

            //添加子属性
            m_jobArguments->addAttr(INPUT_FILE_NAMES, m_inputFileNames);
            m_jobArguments->addAttr(OUTPUT_FILE_NAMES, m_outputFileNames);
            m_jobArguments->addAttr(PARAMETER_ARGS, m_parameterArgs);
        }
    }
```

```cpp
public:
    //获取属性:输入文件名列表
    virtual orsIProperty *getInputFileNames()
    {
        return m_inputFileNames.get();
    }

    //获取属性:算法参数列表
    virtual orsIProperty *getParameterArgs()
    {
        return m_parameterArgs.get();
    }

    //获取属性:输出文件列表
    virtual orsIProperty *getOutputFileNames()
    {
        return m_outputFileNames.get();
    }

    //获取对象属性:重载orsIObject方法
    virtual const orsIProperty *getProperty() const
    {
        return m_jobArguments.get();
    }

    //利用属性初始化对象: 重载orsIObject方法
    virtual bool initFromProperty(orsIProperty *property)
    {
        ref_ptr<orsIProperty> inputFileNames;
        ref_ptr<orsIProperty> parameterArgs;
        ref_ptr<orsIProperty> outputFileNames;

        //提取子属性:输入文件名列表
        property->getAttr(INPUT_FILE_NAMES, inputFileNames);
        //提取子属性:算法参数列表
        property->getAttr(PARAMETER_ARGS, parameterArgs);
        //提取子属性:输入文件名列表
```

```cpp
        property->getAttr(OUTPUT_FILE_NAMES, outputFileNames);

    //兼容旧的调用方式
    if(NULL == inputFileNames.get())
    inputFileNames = property;
    if(NULL == parameterArgs.get())
    parameterArgs =  property;
    if(NULL == outputFileNames.get())
    outputFileNames = property;

    //自动更新内部属性
    if(inputFileNames.get() != m_inputFileNames.get())
        m_inputFileNames->update(inputFileNames.get());
    if(outputFileNames.get() != m_outputFileNames.get())
        m_outputFileNames->update(outputFileNames.get());
    if(parameterArgs.get() != m_parameterArgs.get())
        m_parameterArgs->update(parameterArgs.get());

    //设置标准可执行对象参数
    return this->setArguments(inputFileNames.get(), parameterArgs.get(),
        outputFileNames.get());
}

protected:
    ref_ptr<orsIProperty> m_jobArguments;
    ref_ptr<orsIProperty> m_inputFileNames;
    ref_ptr<orsIProperty> m_parameterArgs;
    ref_ptr<orsIProperty> m_outputFileNames;
};
```

2. 栅格处理可执行对象帮助模板

orsIRasterExeHelper 用于栅格处理。它在普通可执行对象帮助模板的基础上增加了 ROI 属性。有了 ROI 属性,可以指定处理的区域用于影像局部处理。在此基础上,可以统一地实现栅格处理的并行,而不需要对每个对象进行并行编程。

```cpp
template <typename orsIExeInteface>
class orsIRasterExeHelper: public orsIExeInteface
{
    protected:
        orsIRasterExeHelper(bool bForRegister)
        {
            //对象注册?
            if(!bForRegister) {
                //创建可执行对象属性:作业参数
                m_jobArguments = getPlatform()->createProperty();
                //创建子属性:输入文件名列表
                m_inputFileNames = getPlatform()->createProperty();
                //创建子属性:算法参数列表
                m_parameterArgs = getPlatform()->createProperty();
                //创建子属性:输出文件名列表
                m_outputFileNames = getPlatform()->createProperty();
                //创建子属性:ROI参数列表
                m_roiArgs = getPlatform()->createProperty();

                //添加子属性
                m_jobArguments->addAttr(INPUT_FILE_NAMES, m_inputFileNames);
                m_jobArguments->addAttr(OUTPUT_FILE_NAMES, m_outputFileNames);
                m_jobArguments->addAttr(PARAMETER_ARGS, m_parameterArgs);

                m_jobArguments->addAttr(ORS_RASTER_ROI, m_roiArgs);
            }
        }

    public:
        //获取属性:输入文件名列表
        virtual orsIProperty * getInputFileNames()
        {
            return m_inputFileNames.get();
        }

        //获取属性:算法参数列表
        virtual orsIProperty * getParameterArgs()
```

```cpp
{
    return m_parameterArgs.get();
}

//获取属性:输出文件列表
virtual orsIProperty *getOutputFileNames()
{
    return m_outputFileNames.get();
}

//获取对象属性:重载 orsIObject 方法
virtual const orsIProperty *getProperty() const
{
    return m_jobArguments.get();
}

//ROI 设置接口
virtual bool setROI(orsIProperty *roiArgs) = 0;

//利用属性初始化对象: 重载 orsIObject 方法
virtual bool initFromProperty(orsIProperty *property)
{
    ref_ptr<orsIProperty> inputFileNames;
    ref_ptr<orsIProperty> parameterArgs;
    ref_ptr<orsIProperty> outputFileNames;
    ref_ptr<orsIProperty> roiArgs;

    //提取子属性:输入文件名列表
    property->getAttr(INPUT_FILE_NAMES, inputFileNames);
    //提取子属性:算法参数列表
    property->getAttr(PARAMETER_ARGS, parameterArgs);
    //提取子属性:输入文件名列表
    property->getAttr(OUTPUT_FILE_NAMES, outputFileNames);
    //提取子属性:ROI 参数列表
    property->getAttr(ORS_RASTER_ROI, roiArgs);

    //兼容旧的调用方式
    if(NULL == inputFileNames.get())
```

```cpp
            inputFileNames = property;
        if(NULL == parameterArgs.get())
            parameterArgs = property;
        if(NULL == outputFileNames.get())
            outputFileNames = property;
        if(NULL == roiArgs.get())
            roiArgs = property;

        //自动更新内部属性
        if(inputFileNames.get() != m_inputFileNames.get())
            m_inputFileNames->update(inputFileNames.get());
        if(outputFileNames.get() != m_outputFileNames.get())
            m_outputFileNames->update(outputFileNames.get());
        if(parameterArgs.get() != m_parameterArgs.get())
            m_parameterArgs->update(parameterArgs.get());

        if(roiArgs.get() != m_roiArgs.get())
            roiArgs->update(roiArgs.get());
        //设置 ROI
        this->setROI(roiArgs.get());

        //设置标准可执行对象参数
        return this->setArguments(inputFileNames.get(), parameterArgs.get(), outputFileNames.get());
    }

protected:
    ref_ptr<orsIProperty> m_jobArguments;

    ref_ptr<orsIProperty> m_inputFileNames;
    ref_ptr<orsIProperty> m_parameterArgs;
    ref_ptr<orsIProperty> m_outputFileNames;

    ref_ptr<orsIProperty> m_roiArgs;
};
```

10.2.4 实例——中值滤波

基于 OpenRS 可执行对象,可实现单任务中值滤波处理和并行任务处理。为了说明可执行对象的实现灵活性,文件读写使用了原始的 GDAL API。

1. 滤波对象接口

单任务滤波接口:

```
interface orsISE_ImageFilter : public orsISimpleExe
{
public:
    ORS_INTERFACE_DEF(orsISimpleExe, "imageFilter");
};
```

并行任务滤波接口:

```
interface orsIPE_ImageFilter : public orsIParallelExe
{
public:
    ORS_INTERFACE_DEF(orsIParallelExe, "imageFilter");
};
```

2. 单任务中值滤波实现

为了支持不同数值类型的影像滤波,定义中值滤波模板如下:

```
//中值滤波算法接口
interface orsIMedianFilter
{
    virtual bool medianFilter(const void * inBuf, int bufWid, int bufHei, void * out-
        Buf, orsIProcessMsg * msg, float fromPercent, float totalPercent) = 0;
};
//中值滤波模板
template<class _T>
class orsXMedianFilter : public orsIMedianFilter
{
private:
    void bubbleSort(_T * buf, int n)
```

```cpp
    {
        for(int j = n - 1; j > 0; j--) //外层循环
        {
            for(int i = 0; i < j; i++) //内层循环
            {
                if(buf[i] > buf[i+1]) //变换
                {
                    buf[i] = buf[i] + buf[i+1];
                    buf[i+1] = buf[i] - buf[i+1];
                    buf[i] = buf[i] - buf[i+1];
                }
            }
        }
    }

public:
    virtual bool medianFilter(const void * inBuf, int bufWid, int bufHei, void * out-
        Buf0, orsIProcessMsg * msg, float fromPercent, float totalPercent)
    {
        _T w[9];
        const _T * buf0, * buf1, * buf2;
        buf0 = (_T *)inBuf;
        buf1 = buf0 + bufWid;
        buf2 = buf1 + bufWid;

        _T * outBuf = (_T *)outBuf0;
        //复制第一行和最后一行
        memcpy(outBuf, buf0, bufWid * sizeof(_T));
        memcpy(outBuf+(bufHei-1)*bufWid, buf0+(bufHei-1)*bufWid, bufWid *
            sizeof(_T));
        outBuf += bufWid;

        long progress = 0, progrss1;
        for(int i=0; i< bufHei-2; i++)
        {
            //第一个像素
            *outBuf++ = *buf1;
            for(int j=0; j< bufWid-2; j++)
```

```cpp
        {
            w[0] = *buf0++;w[1] = *buf0;w[2] = buf0[1];
            w[3] = *buf1++;w[4] = *buf1;w[5] = buf1[1];
            w[6] = *buf2++;w[7] = *buf2;w[8] = buf2[1];

            bubbleSort(w, 9);

            *outBuf++ = w[4];
        }
        //最后一个像素
        *outBuf++ = buf1[1];
        buf0+=2;buf1+=2;buf2+=2;

        //进度设置,只有当进度超过了1%才进行更新
        progrss1 = 100 * i/bufHei;
        if(progrss1 > progress) {
            if(!msg->process(fromPercent + 0.01 * progrss1 * totalPercent))
                return false;

            progress = progrss1;
        }
    }
    return true;
    }
};
//单任务中值滤波实现类
class orsSE_MedianFilter:public orsIRasterExeHelper<orsISE_ImageFilter>, public
    orsObjectBase
{
private:
    orsStringm_inputImageFileName;//输入文件名
    orsStringm_outputImageFileName;//输出文件名

    //ROI
    ors_int32m_fromRow;//起始行
    ors_int32m_toRow;//结束行


public:
```

```cpp
    orsSE_MedianFilter(bool bForRegister);
    virtual ~orsSE_MedianFilter();

    virtual bool setROI(orsIProperty * roiArgs);

    virtual bool setArguments(orsIProperty * inputFileNames, orsIProperty * parame-
        terArgs, orsIProperty * outputFileNames);

    virtual ref_ptr<orsIProperty> execute(orsIProcessMsg * process);

    ORS_OBJECT_IMP3(orsSE_MedianFilter, orsISE_ImageFilter, orsISimpleExe,
        orsIExecute, _T("median"), _T("Simple Median Filter"))
};
//构造函数
orsSE_MedianFilter::orsSE_MedianFilter(bool bForRegister)
    : orsIRasterExeHelper<orsISE_ImageFilter>(bForRegister)
{
    //对象注册?
    if(!bForRegister) {
        m_inputImageFileName = "Input Image File";
        m_outputImageFileName = "Output Image File";

        //都为零,代表整个影像
        m_fromRow = m_toRow = 0;

        //输入文件属性
        m_inputFileNames->addAttr("inputImageFile", m_inputImageFileName);
        //输出文件属性
        m_outputFileNames->addAttr("outputImageFile", m_outputImageFileName);

        //ROI 属性:起始行
        m_roiArgs->addAttr(ORS_RROI_BEGINROW, m_fromRow);
        //ROI 属性:行数
        m_roiArgs->addAttr(ORS_RROI_NUMOFROWS, m_toRow-m_fromRow);
    }
}

orsSE_MedianFilter::~orsSE_MedianFilter()
```

```cpp
{
}

//输入文件名、算法参数、输出文件名设置
bool orsSE_MedianFilter::setArguments(orsIProperty * inputFileNames, orsIProperty
    * parameterArgs, orsIProperty * outputFileNames)
{
    inputFileNames->getAttr("inputImageFile", m_inputImageFileName);
    outputFileNames->getAttr("outputImageFile", m_outputImageFileName);
    parameterArgs->getAttr("fromRow", m_fromRow);
    parameterArgs->getAttr("toRow", m_toRow);

    return true;
}

//ROI 设置
bool orsSE_MedianFilter::setROI(orsIProperty * roiArgs)
{
    roiArgs->getAttr(ORS_RROI_BEGINROW, m_fromRow);
    roiArgs->getAttr(ORS_RROI_NUMOFROWS, m_toRow);
    m_toRow += m_fromRow;

    return true;
}

#include "GDALImageWriter.h"

//对象执行
ref_ptr<orsIProperty> orsSE_MedianFilter::execute(orsIProcessMsg * msg)
{
    //打开影像
    GDALDatasetH hDataset = GDALOpen(m_inputImageFileName, GA_ReadOnly);
    if(hDataset == NULL)
    {
        getPlatform()->logPrint(ORS_LOG_ERROR, "GDALOpen failed — %d\n%s\n", CPLGetLastErrorNo(), CPLGetLastErrorMsg());
        return NULL;
    }
```

```cpp
//影像大小
long imgWid = GDALGetRasterXSize(hDataset);
long imgHei = GDALGetRasterYSize(hDataset);
long numOfBands = GDALGetRasterCount(hDataset);

//数值类型
GDALDataType dataType = GDALGetRasterDataType(GDALGetRasterBand(hDataset, 1));

geoImageINFO imgInfo;
CGDALImageWriter *pImgWriter;

//ROI 没有设置,设置为整个影像
if(m_toRow <= m_fromRow)
    m_toRow = imgHei;
//处理行数
long nRows = m_toRow - m_fromRow;
{
    imgInfo.imgWid = imgWid;
    imgInfo.imgHei = nRows;
    imgInfo.tile_wid = imgWid;
    imgInfo.tile_hei = nRows;

    //子影像在原始影像中的位置
    imgInfo.transType = geoImgTranOFFSETSCALE;
    imgInfo.i0 = 0; imgInfo.x0 = 1e-20;//避免被 GDAL 忽略
    imgInfo.j0 = 0; imgInfo.y0 = m_fromRow;

    imgInfo.xScale = 1;
    imgInfo.yScale = 1;

    imgInfo.nBands = numOfBands;
    imgInfo.dataType = (orsDataTYPE) dataType;

    //创建输出影像
    pImgWriter = CreateImageWriter(m_outputImageFileName, imgInfo, false);
    if(pImgWriter == NULL)
    {
```

```cpp
            msg->logPrint(ORS_LOG_ERROR, "Failed creating %s", m_outputImage-
                FileName);
        }
    }

    orsIMedianFilter *filter;

    //根据数值类型创建该类型的中值滤波算法实例
    switch(dataType) {
    case GDT_Byte:
        filter = new orsXMedianFilter<ors_uint8>;
        break;
    case GDT_UInt16:
        filter = new orsXMedianFilter<ors_uint16>;
        break;
    case GDT_Int16:
        filter = new orsXMedianFilter<ors_int16>;
        break;
    case GDT_UInt32:
        filter = new orsXMedianFilter<ors_uint32>;
        break;
    case GDT_Int32:
        filter = new orsXMedianFilter<ors_int32>;
        break;
    case GDT_Float32:
        filter = new orsXMedianFilter<ors_float32>;
        break;
    case GDT_Float64:
        filter = new orsXMedianFilter<ors_float64>;
        break;
    default:
        CPLAssert(FALSE);
        return 0;
    }

    //分段数,每一段处理 64M 像素
    int nSegRows = 64L * 1024 * 1024 / imgWid;
```

```
int bandPixelBytes = GDALGetDataTypeSize(dataType) / 8;
long nBandBytes = imgWid * nSegRows * bandPixelBytes;

//分配单个波段处理所需的内存
BYTE * inBuf = (BYTE *)malloc(nBandBytes);
BYTE * outBuf = (BYTE *)malloc(nBandBytes);

int iRow;
float fromPercent, totalPercent;
fromPercent = 0;

//分段处理
for(iRow = m_fromRow; iRow < m_toRow; iRow += nSegRows)
{
    if(iRow + nSegRows <= m_toRow)
        nRows = nSegRows;
    else
        nRows = m_toRow - iRow;

    totalPercent = (float)nRows / ((m_toRow - m_fromRow) * numOfBands);

    int iBand;
    //分波段处理
    for(iBand = 0; iBand < numOfBands; iBand++)
    {
        //读取该波段影像
        GDALRasterIO(GDALGetRasterBand(hDataset, iBand+1), GF_Read,
            0, iRow, imgWid, nRows, inBuf, imgWid, nRows,
            dataType, bandPixelBytes, imgWid * bandPixelBytes);

        //对波段影像进行滤波
        if(!filter->medianFilter(inBuf, imgWid, nRows, outBuf, msg, fromPercent,
            totalPercent)) {
            break;
        }

        //写该波段滤波后数据
        pImgWriter->WriteBandRect(iBand, 0, iRow-m_fromRow, imgWid, nRows, out-
```

```
            Buf, imgWid);

        fromPercent += totalPercent;
    }
}
//关闭输入文件
GDALClose(hDataset);

//关闭输出文件
if(pImgWriter)
    delete pImgWriter;

//释放缓冲区
free(inBuf);
free(outBuf);
ref_ptr<orsIProperty> jobOutput = getPlatform()->createProperty();

jobOutput->addAttr("out:fileName", m_outputImageFileName);

return jobOutput;
}
```

3. 并行任务中值滤波实现

由于 orsSE_MedianFilter 中定义 ROI 属性,因此 orsSE_MedianFilter 可以自动并行化。下面给出的代码只是为了展示并行处理对象的编写方法。

```
//分布式并行中值滤波实现类
class orsPE_MedianFilter : public orsIExeHelper<orsIPE_ImageFilter>, public orsObjectBase
{
private:
    orsString m_inputImageFileName;
    orsString m_outputImageFileName;

public:
    orsPE_MedianFilter(bool bForRegister);
    virtual ~orsPE_MedianFilter();
```

```cpp
    virtual bool setArguments(orsIProperty * inputFileNames, orsIProperty * parame-
        terArgs, orsIProperty * outputFileNames);

    //任务分割方法
    virtual orsArray<ref_ptr<orsIProperty>> getTasks(int nTasks = 0);
    //任务执行方法
    virtual ref_ptr<orsIProperty> taskExecute(ref_ptr<orsIProperty> taskInput, or-
        sIProcessMsg * msg);
    //任务合并方法
    virtual ref_ptr<orsIProperty> taskCombine(orsArray<ref_ptr<orsIProperty>>
        taskInputs, orsIProcessMsg * msg);

    ORS_OBJECT_IMP3(orsPE_MedianFilter, orsIPE_ImageFilter, orsIParallelExe, ors-
        IExecute, _T("medianFiler"), _T("Distributed Median Filter"))
};

//构造函数
orsPE_MedianFilter::orsPE_MedianFilter(bool bForRegister)
    :orsIExeHelper<orsIPE_ImageFilter>(bForRegister)
{
    //对象注册?
    if(!bForRegister) {
        m_inputImageFileName = "Input Image File";
        m_outputImageFileName = "Output Image File";

        //输入文件属性
        m_inputFileNames->addAttr("inputImageFile", m_inputImageFileName);
        //输出文件属性
        m_outputFileNames->addAttr("outputImageFile", m_outputImageFileName);
    }
}

orsPE_MedianFilter::~orsPE_MedianFilter()
{
}

//可执行对象输入文件、算法参数、输出文件设置
```

```cpp
bool orsPE_MedianFilter::setArguments(orsIProperty * inputFileNames, orsIProperty
    * parameterArgs, orsIProperty * outputFileNames)
{
    inputFileNames->getAttr("inputImageFile", m_inputImageFileName);
    outputFileNames->getAttr("outputImageFile", m_outputImageFileName);

    return true;
}

//任务分割
orsArray<ref_ptr<orsIProperty> > orsPE_MedianFilter::getTasks(int nTasks)
{
    orsArray<ref_ptr<orsIProperty> > tasks;

    //获取影像信息
    GDALDatasetH hDataset = GDALOpen(m_inputImageFileName, GA_ReadOnly);
    if(hDataset == NULL)
    {
        getPlatform()->logPrint(ORS_LOG_ERROR, "GDALOpen failed - %d\n%s\n",
            CPLGetLastErrorNo(), CPLGetLastErrorMsg());
        return tasks;
    }

    long imgWid = GDALGetRasterXSize(hDataset);
    long imgHei = GDALGetRasterYSize(hDataset);

    GDALClose(hDataset);
    //若没有指定任务数,则默认任务数为8
    if(0 == nTasks)
        nTasks = 8;

    int fromRow = 0;
    //每个子任务的行数
    int subRows = (imgHei + nTasks - 1) / nTasks;

    //设置任务参数
    for(int i=0; i< nTasks; i++)
    {
```

```cpp
        //建立任务参数列表
        ref_ptr<orsIProperty> task = getPlatform()->createProperty();

        //最后一个任务,调整为剩下的函数
        if(i == nTasks-1)
            subRows = imgHei - fromRow;

        //子任务起始行
        task->addAttr("task:fromRow", (ors_int32)fromRow);
        //子任务结束行
        task->addAttr("task:toRow", (ors_int32)(fromRow + subRows));
        tasks.push_back(task);

        fromRow += subRows;
    }

    return tasks;
}

#include "orsPE_MedianFilter.h"
//子任务执行
ref_ptr<orsIProperty> orsPE_MedianFilter::taskExecute(ref_ptr<orsIProperty>
    taskInput, orsIProcessMsg* msg)
{
    msg->process(0.0);
    msg->logPrint(ORS_LOG_INFO,"task begin");

    ors_int32 fromRow;
    ors_int32 toRow;

    //子任务起始行
    taskInput->getAttr("task:fromRow", fromRow);
    //子任务结束行
    taskInput->getAttr("task:toRow", toRow);

    //子任务结果输出文件

    orsString taskfile_out;
```

```cpp
taskfile_out = m_outputImageFileName;
int pos = taskfile_out.reverseFind('.');

ref_ptr<orsIProperty> task = getPlatform()->createProperty();
if(pos > -1) {
    taskfile_out = taskfile_out.left(pos);

    //形成子任务文件名
    char buf[256];
    sprintf(buf, "%s_%d.tif", taskfile_out.c_str(), fromRow);
    taskfile_out = buf;

    //子任务执行信息
    task->addAttr("task:fromRow", fromRow);
    task->addAttr("task:toRow", toRow);
    task->addAttr("task:file_out", taskfile_out);

    msg->logPrint(ORS_LOG_INFO, "task file out : %s", taskfile_out.c_str());
    //调用 orsSE_MedianFilter 执行子任务

    //子任务执行参数
    ref_ptr<orsIProperty> subJobPara = getPlatform()->createProperty();

    //子任务参数:输入文件名
    subJobPara->addAttr("inputImageFile", m_inputImageFileName.c_str());
    //子任务参数:输出文件名
    subJobPara->addAttr("outputImageFile", taskfile_out.c_str());
    //子任务参数:起始行和行数
    subJobPara->addAttr(ORS_RROI_BEGINROW, fromRow);
    subJobPara->addAttr(ORS_RROI_NUMOFROWS, toRow-fromRow);

    //创建单任务执行对象
    ref_ptr<orsISimpleExe> filterExe =
        ORS_CREATE_OBJECT(orsISimpleExe, ORS_SE_IMAGEFILTER_MEDIAN);

    if(NULL != filterExe.get()) {
        //初始化单任务可执行对象
```

```cpp
        filterExe->initFromProperty(subJobPara.get());
        //执行单任务滤波
        filterExe->execute(msg);
    }

    }
    else{
        getPlatform()->logPrint(ORS_LOG_ERROR, "Invalid output image file name");
    }
    return task;
}

//任务合并
ref_ptr<orsIProperty> orsPE_MedianFilter::taskCombine(orsArray<ref_ptr<orsIProp-
    erty> > taskInputs, orsIProcessMsg * msg)
{
    //创建任务合并信息
    ref_ptr<orsIProperty> jobOut = getPlatform()->createProperty();
    //合并的任务数
    jobOut->addAttr("job:tasknum", (ors_int32)taskInputs.size());
    //合并输出的文件名
    jobOut->addAttr("filtered file", m_outputImageFileName);

    msg->process(0);
    msg->logPrint(ORS_LOG_INFO, "开始合并影像");
    long imgWid, imgHei;
    long numOfBands;
    GDALDataType dataType;

    //获取输入文件信息
    {
        GDALDatasetH hDataset = GDALOpen(m_inputImageFileName, GA_ReadOnly);
        if(hDataset == NULL)
        {
            getPlatform()->logPrint(ORS_LOG_ERROR, "GDALOpen failed - %d\n%s\
                n", CPLGetLastErrorNo(), CPLGetLastErrorMsg());
            return NULL;
```

```
        }

        imgWid = GDALGetRasterXSize(hDataset);
        imgHei = GDALGetRasterYSize(hDataset);
        numOfBands = GDALGetRasterCount(hDataset);

        dataType = GDALGetRasterDataType(GDALGetRasterBand(hDataset, 1));

        GDALClose(hDataset);
    }
    //输出文件信息
    geoImageINFO imgInfo;
    CGDALImageWriter * pImgWriter;

    imgInfo.imgWid = imgWid;
    imgInfo.imgHei = imgHei;
    imgInfo.tile_wid = 256;
    imgInfo.tile_hei = 256;

    //子影像在原始影像中的位置
    imgInfo.transType = geoImgTranOFFSETSCALE;
    imgInfo.i0 = 0; imgInfo.x0 = 0;
    imgInfo.j0 = 0; imgInfo.y0 = 0;

    imgInfo.xScale = 1;
    imgInfo.yScale = 1;

    imgInfo.nBands = numOfBands;
    imgInfo.dataType = (orsDataTYPE) dataType;

    pImgWriter = CreateImageWriter(m_outputImageFileName, imgInfo, false);
    if (pImgWriter == NULL)
    {
        msg->logPrint(ORS_LOG_ERROR, "Open create image writer");
    }

    int bandPixelBytes = GDALGetDataTypeSize(dataType) / 8;
```

```
BYTE * inBuf = NULL;

int iImg;
for(iImg=0; iImg<taskInputs.size(); iImg++)
{
    orsString imageName;
    ors_int32 fromRow, toRow;
    long numOfRows;

    //子任务对应的起始行
    taskInputs[iImg]->getAttr("task:fromRow",fromRow);
    taskInputs[iImg]->getAttr("task:toRow",toRow);

    //子任务对应的结束行
    taskInputs[iImg]->getAttr("task:file_out", imageName);

    //行数
    numOfRows = toRow - fromRow;

    if(NULL == inBuf) {
        long nBandBytes = imgWid * numOfRows * bandPixelBytes;
        inBuf = (BYTE *)malloc(nBandBytes);
    }
    if(NULL != imageName.c_str()) {
        GDALDatasetH hDataset = GDALOpen(imageName, GA_ReadOnly);
        if(hDataset == NULL)
        {
            getPlatform()->logPrint(ORS_LOG_ERROR, "GDALOpen failed - %d\
                n%s\n", CPLGetLastErrorNo(), CPLGetLastErrorMsg());
            continue;
        }

        //按波段读写
        for(int iBand =0; iBand < numOfBands; iBand++)
        {
            GDALRasterIO(GDALGetRasterBand(hDataset, iBand+1), GF_Read, 0, 0,
                imgWid, numOfRows, inBuf, imgWid, numOfRows, dataType, bandPixel-
                Bytes, imgWid * bandPixelBytes);
```

```
            pImgWriter->WriteBandRect(iBand, 0, fromRow, imgWid, numOfRows,
                inBuf, imgWid);
        }
        //合并进度信息
        msg->process((double)iImg / taskInputs.size());
        msg->logPrint(ORS_LOG_INFO,"image %d has been merged", iImg);

        GDALClose(hDataset);
    }
}

    free(inBuf);

    delete pImgWriter;

    return jobOut;
}
```

10.3 细粒度插件(依赖于 OpenRS 遥感处理对象体系)

在 OpenRS 中,算法对象(orsIAlgorithm)和影像源对象(orsIImageSource)是两类重要的对象。

10.3.1 算法对象

orsIAlgorithm 是纯粹的算法对象的接口,包含数据的读取、存储等功能,是其他对象,如影像源对象的实现基础。把算法对象挂接到相应类别的影像源,就可以使用同一影像源对象实现不同算法的调用。

10.3.2 影像链节点——影像源对象

影像处理链的基本要求就是要加入影像链中的处理节点必须实现影像源接口(ImageSource)。只要实现影像源接口就可以作为节点加入影像链(图 10-2)。

10.3.3 从算法到可执行对象

如图 10-3 所示,算法、影像链与可执行对象的关系是一个包含关系,即影像链的节点处理要用到算法,而可执行对象可以在影像链基础上通过增加输入输出文

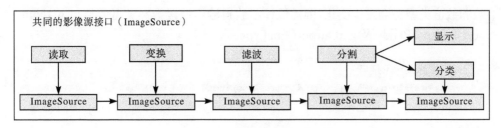

图 10-2　影像链的实现原理

件的打开和关闭来实现。在可执行对象上增加 ROI 属性，又可以支持并行处理。这样就可以有效地进行代码重用，提高编程的效率和软件的可靠性。

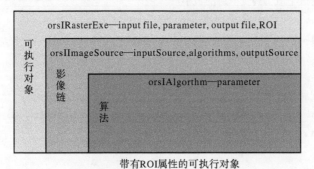

图 10-3　从算法到可执行对象的属性关系图

1. orsIAlgorithm 对象

算法对象接口只起到一个接口分类的作用。

```
interface orsIAlgorithm: public orsIObject
{
public:
    ORS_INTERFACE_DEF(orsIObject, _T("algorithm"))
};
```

不同类别的算法因为算法的目的不同，输入输出也不一样，因此需要针对不同类别的算法，分别定义不同的算法接口。其中，栅格类算法对象的输入输出为影像块。下面以影像滤波算法接口和 Lee 滤波算法对象为例进行说明。

影像滤波算法接口以影像块 orsIImageData 为输入和输出，因为滤波要考虑像素邻域，所以输入的影像窗口往往要比输出的大，因此设计了 getExtendRect 接口用于获取滤波算法接口需要扩展的输入影像范围。有时滤波之后的影像波段数和数值类型都会发生变化,因此设计了获得输出波段数和类型的接口 getNumberOfOutputBands 及 getOutputDataType。

```cpp
//影像滤波对象接口
class orsIAlgImageFilter : public orsIAlgorithm
{
public:
    //根据一个矩形框得到一个扩展的矩形框,调用者将输入这个扩展的矩形框的数据
    virtual orsRect_i getExtendRect(orsRect_i &proceedRect) = 0;

    //通过最小值得到变换后的最小值
    virtual ors_float64 getNullSampleValue(ors_uint band) = 0;
    virtual ors_float64 getMinSampleValue(ors_uint band) = 0;
    virtual ors_float64 getMaxSampleValue(ors_uint band) = 0;

    //处理后的波段数
    virtual ors_uint getNumberOfOutputBands(ors_uint inputBandNum) = 0;
    virtual orsDataTYPE getOutputDataType() = 0;

    //实际处理
    virtual void filter(const orsIImageData * src, orsIImageData * dst) = 0;

    ORS_INTERFACE_DEF(orsIAlgorithm, _T("imageFilter"))
};
```

Lee 滤波算法对象实现滤波接口,并定义自己特有的属性参数,具体细节请参看源代码\OpenRS\desktop\src\plugins 目录下的 ors_ZG_SARFilter.h 和 ors_ZG_SARFilter.cpp。

```cpp
//Lee 滤波对象
class orsAlgImageFilterLee: public orsIAlgImageFilter, public orsObjectBase
{
public:
    orsAlgImageFilterLee();
    ~orsAlgImageFilterLee();

    //得到 detailed description
    virtual ors_stringgetDesc() const { return "SAR Lee 滤波"; }
```

```cpp
    orsRect_i getExtendRect(orsRect_i &proceedRect);
    ors_float64 getNullSampleValue(ors_uint band);
    ors_float64 getMinSampleValue(ors_uint band);
    ors_float64 getMaxSampleValue(ors_uint band);
    ors_uint getNumberOfOutputBands(ors_uint inputBandNum);
    orsDataTYPE getOutputDataType();
    void filter(const orsIImageData * src,orsIImageData * dst);

    virtual const orsIProperty * getProperty() const;
    virtual bool initFromProperty(orsIProperty * property);

private:
    template<class T>
    void filter(T, const orsIImageData * src,orsIImageData * dst);

    const orsIImageData * m_srcImageData;
    orsIImageData * m_dstImageData;

    //算法参数属性
    ref_ptr<orsIProperty> m_inputProperty;

    ors_int32 m_nLooks;//视数
    ors_int32 m_nFilterWidX;//滤波窗口 X 方向大小
    ors_int32 m_nFilterWidY;//滤波窗口 Y 方向大小

    ORS_OBJECT_IMP1(orsAlgImageFilterLee,orsIAlgImageFilter,"Lee", "Lee Filter")
};
```

2. orsIImageSource 对象

滤波类影像源也只是一个分类的名字。

```cpp
interface orsIImageSourceFilter: public orsIImageSource
{
    virtual void setFilterAlgortithm(orsIAlgImageFilter * filter) = 0;
    ORS_INTERFACE_DEF(orsIImageSource, _T("filter"))
};
```

和滤波算法不同,滤波影像源对象不需要针对每个算法写一个,只需要统一地写一个而把算法对象通过 setFilterAlgortithm 设置。具体细节请参看源代码\OpenRS\desktop\src\orsImage 目录下的 orsXImageSourceFilter.h 和 orsXImageSourceFilter.cpp。

```
class orsXImageSourceFilter : public orsIImageSourceHelper_prop<orsIImageSource-
    Filter>, public orsObjectBase
{
public:
    orsXImageSourceFilter(bool bForRegister);
    virtual bool setArguments(orsIProperty * inputSourcePtrs, orsIProperty * param-
        eterArgs, orsIProperty * outputSourcePtrs);

    //滤波算法设置
    virtual void setFilterAlgortithm(orsIAlgImageFilter * filter)
    {
        m_algFilter = filter;
    }

    //获取滤波后的数据
    virtual orsIImageData * getImageData(orsRect_i &rect, double zoomRate, orsBand-
        Set &bandSet);

protected:
    //滤波算法对象指针
    ref_ptr<orsIAlgImageFilter>m_algFilter;
    ref_ptr<orsIImageData>m_imageData;
    ...
};
```

3. orsIImageSource 对象自动包装为 orsISimpleExe 对象

ImageSource 对象可以通过模板 orsIExeImageSourceHelper 和宏 ORS_EXE_IMP 自动包装为带 ROI 参数的 orsIExe 对象。该模板在 orsIExeImageSourceHelper.h 文件中。

orsIExeImageSourceHelper 模板定义：

```
//该帮助类用于将一个实现了 orsIImageSource 接口的对象自动转为一个 orsISimpleExe
//对象
```

```cpp
template<class _t = orsISimpleExe>
class orsIExeImageSourceHelper : public _t
{
public:
    virtual ~orsIExeImageSourceHelper()
    {
    }

    //描述
    bool setDescribe(const orsString &describe)
    {
        m_describe = describe;
    }

    virtual ors_stringgetDesc() const
    {
        return m_describe;
    }

    //输入、输出、算法参数
    virtual bool setArguments(orsIProperty * inputFileNames, orsIProperty * parame-
        terArgs, orsIProperty * outputFileNames)
    {
        ors_uint32 index;
        ors_string name;
        orsVariantType type;
        ors_int32 numOfValues;
        orsString fileName;

        m_strInPathNames.clear();

        //不止一个输入数据源,如影像融合
        if(m_inputSourcePtrs->size() > 0){
            for( int i=0; i<m_inputSourcePtrs->size(); i++)
            {
                //从输入数据源获取输入文件属性名
                m_inputSourcePtrs->getAttributeInfo(i, name, type, numOfValues);
                inputFileNames->getAttr(name, fileName);
```

```cpp
            m_strInPathNames.push_back(fileName);
        }
    }
    else{    //只有一个输入数据源,缺省名字
        m_inputFileNames->getAttr("InputImageFileName", fileName);
        m_strInPathNames.push_back(fileName);
    }
    //不止一个输出数据源
    if(m_outputSourcePtrs->size() > 0){
        for(int i=0; i<m_outputSourcePtrs->size(); i++)
        {
            m_outputSourcePtrs->getAttributeInfo(i, name, type, numOfValues);
            m_outputFileNames->getAttr(name, fileName);
            m_strOutPathNames.push_back(fileName);
        }
    }
    else{    //只有一个输出数据源,缺省名字
        m_outputFileNames->getAttr("OutputImageFileName", fileName);
        m_strOutPathNames.push_back(fileName);
    }

    //初始化被包装的 orsIImageSource 对象
    m_pAlgSource->initFromProperty(m_sourceArgs.get());
    return true;
}


bool setROI(orsIProperty *roiArgs)
{
    m_temRectWidth = 256;
    m_temRectHeight = 256;

    roiArgs->getAttr(ORS_RROI_BEGINROW, m_boundingRect.m_xmin);
    roiArgs->getAttr(ORS_RROI_NUMOFROWS, m_boundingRect.m_xmax);
    m_boundingRect.m_xmax += m_boundingRect.m_xmin;

    roiArgs->getAttr(ORS_RROI_BEGINCOL, m_boundingRect.m_ymin);
    roiArgs->getAttr(ORS_RROI_NUMOFCOLS, m_boundingRect.m_ymax);
```

```cpp
            m_roiArgs->getAttr("Write_tileWidth",m_temRectWidth);
            m_roiArgs->getAttr("Write_tileHeight",m_temRectHeight);

            m_boundingRect.m_ymax += m_boundingRect.m_ymin;

            return true;
        }

    protected:
        orsIExeImageSourceHelper(bool bForRegister)
        {
        }

        void create(ref_ptr<orsIImageSource> algSource)
        {
            m_pAlgSource = algSource;

            m_jobArguments = getPlatform()->createProperty();
            m_inputFileNames = getPlatform()->createProperty();
            m_outputFileNames = getPlatform()->createProperty();
            m_roiArgs = getPlatform()->createProperty();
            m_outputMsg = getPlatform()->createProperty();
            m_outputMsg->addAttr("system__","1");
            {
                ors_string name;
                orsVariantType type;
                ors_int32 numOfValues;

                m_sourceArgs = algSource->getProperty();

                assert(NULL != m_sourceArgs.get());

                m_sourceArgs->getAttr(ORS_SOURCES_INPUT, m_inputSourcePtrs);
                m_sourceArgs->getAttr(ORS_SOURCES_OUTPUT, m_outputSourcePtrs);
```

```cpp
if(m_inputSourcePtrs.get() != NULL && m_inputSourcePtrs->size() >
    0){
        for(int i=0; i<m_inputSourcePtrs->size(); i++)
        {
            m_inputSourcePtrs->getAttributeInfo(i,name,type,numOf-
                Values);
            m_inputFileNames->addAttr(name, name);
        }
}
else{//缺省名字
    m_inputFileNames->addAttr("InputImageFileName", "");
}

if(m_outputSourcePtrs.get()!=NULL && m_outputSourcePtrs->size()>
    0){
        for(int i=0; i<m_outputSourcePtrs->size(); i++)
        {
            m_outputSourcePtrs->getAttributeInfo(i,name,type,numO-
                fValues);
            m_outputFileNames->addAttr(name, name);
        }
}
else{//缺省名字
    m_outputFileNames->addAttr("OutputImageFileName", "");
}
}

m_jobArguments->addAttr(INPUT_FILE_NAMES,m_inputFileNames);
m_jobArguments->addAttr(OUTPUT_FILE_NAMES,m_outputFileNames);

m_sourceArgs->getAttr(PARAMETER_ARGS, m_parameterArgs);

if(m_parameterArgs == NULL)
    m_parameterArgs = getPlatform()->createProperty();

m_jobArguments->addAttr(PARAMETER_ARGS, m_parameterArgs);
```

```cpp
        m_temRectWidth = 256;
        m_temRectHeight = 256;

        m_boundingRect.m_xmin = 0;
        m_boundingRect.m_ymin = 0;
        m_boundingRect.m_xmax = 0;
        m_boundingRect.m_ymax = 0;

        m_roiArgs->addAttr(ORS_RROI_BEGINROW, m_boundingRect.m_xmin);
        m_roiArgs->addAttr(ORS_RROI_NUMOFROWS, m_boundingRect.m_xmax - m_bound-
            ingRect.m_xmin);
        m_roiArgs->addAttr(ORS_RROI_BEGINCOL, m_boundingRect.m_ymin);
        m_roiArgs->addAttr(ORS_RROI_NUMOFCOLS, m_boundingRect.m_ymax - m_bound-
            ingRect.m_ymin);

        m_roiArgs->addAttr("Write_tileWidth", m_temRectWidth);
        m_roiArgs->addAttr("Write_tileHeight", m_temRectHeight);

        m_jobArguments->addAttr(ORS_RASTER_ROI, m_roiArgs);
    }

public:
    virtual orsIProperty *getInputFileNames()
    {
        return m_inputFileNames.get();
    }

    //内部参数
    virtual orsIProperty *getParameterArgs()
    {
        return m_parameterArgs.get();
    }

    //输出文件
    virtual orsIProperty *getOutputFileNames()
    {
        return m_outputFileNames.get();
```

```cpp
}

virtual const orsIProperty *getProperty() const
{
    return m_jobArguments.get();
}

//输入参数信息
virtual bool initFromProperty(orsIProperty *property)
{
    ref_ptr<orsIProperty> inputFileNames;
    ref_ptr<orsIProperty> parameterArgs;
    ref_ptr<orsIProperty> outputFileNames;
    ref_ptr<orsIProperty> roiArgs;

    property->getAttr(INPUT_FILE_NAMES, inputFileNames);
    property->getAttr(PARAMETER_ARGS, parameterArgs);
    property->getAttr(OUTPUT_FILE_NAMES, outputFileNames);
    property->getAttr(ORS_RASTER_ROI, roiArgs);

    //兼容旧的调用方式
    if(NULL == inputFileNames.get())
        inputFileNames = property;
    if(NULL == parameterArgs.get())
        parameterArgs = property;
    if(NULL == outputFileNames.get())
        outputFileNames = property;
    if(NULL == roiArgs.get())
        roiArgs = property;

    //自动更新内部属性
    if(inputFileNames.get() != m_inputFileNames.get())
        m_inputFileNames->update(inputFileNames.get());

    if(outputFileNames.get() != m_outputFileNames.get())
        m_outputFileNames->update(outputFileNames.get());
```

```cpp
    if(parameterArgs.get() != m_parameterArgs.get())
        m_parameterArgs->update(parameterArgs.get());

    if(roiArgs.get() != m_roiArgs.get())
        m_roiArgs->update(roiArgs.get());

    setROI(roiArgs.get());

    return setArguments(inputFileNames.get(), parameterArgs.get(), outputFile-
        Names.get());
}

virtual ref_ptr<orsIProperty> execute(orsIProcessMsg * process)
{
    if (process)
    {
        process->logPrint(ORS_LOG_INFO,"translate begin ");
        process->process(0.00);
    }

    orsIPlatform * pPlatform = getPlatform();

    ref_ptr<orsIImageSourceWriter> pWriter = ORS_CREATE_OBJECT(orsIImage-
        SourceWriter,ORS_IMAGESOURCE_WRITER_DEFAULT);

    if(NULL == pWriter.get())
    {
        process->logPrint(ORS_LOG_ERROR, "Can't create writer %s", ORS_IMAGE-
            SOURCE_WRITER_DEFAULT);
        return m_outputMsg.get();
    }

    pWriter->initFromProperty(m_roiArgs.get());

    ref_ptr<orsIImageService> pImageService = (orsIImageService * )getPlat-
        form()->getService(ORS_SERVICE_IMAGE);
    if(pImageService == NULL){
```

```
process->logPrint(ORS_LOG_ERROR,"image service isn't exist");
    return m_outputMsg.get();
}


ref_ptr<orsIImageChain> pImageChain = pImageService->CreateIm-
    ageChain();

orsArray<ref_ptr<orsIImageSourceReader> > pReaders;

int i;
for(i=0; i<m_strInPathNames.size(); i++)
{
    orsIImageSourceReader * pReader;

    pReader = pImageService->openImageFile(m_strInPathNames[i]);

    if (NULL == pReader)
    {
        process->logPrint(ORS_LOG_ERROR, "Can't open %s", m_str-In-
            PathNames[i]);
        return m_outputMsg.get();
    }

    pReaders.push_back(pReader);
}

ref_ptr<orsIImageSource> pAlgSource = m_pAlgSource;//ORS_CREATE_OBJECT
    //(orsIImageSource, m_algID);
if(pAlgSource == NULL)
{
    process->logPrint(ORS_LOG_ERROR,"alg object can't create");
    return m_outputMsg.get();
}

pWriter->SetFileName(m_strOutPathNames[0]);

pAlgSource->disconnectAll();
```

```cpp
        if(1 == pReaders.size()){
            //链接影像链
            pImageChain->add(pReaders[0].get());
            pImageChain->add(pAlgSource.get());
            pImageChain->add(pWriter.get());
        }
        else{
            for(i=0; i< pReaders.size(); i++)
                pAlgSource->connect(pReaders[i].get());

            pWriter->connect(pAlgSource.get());
        }

        //最后结果输出
        pWriter->Write(process);

        return m_outputMsg.get();
    }

protected:

    //source 属性
    ref_ptr<orsIProperty> m_sourceArgs;

    //输入源名字与对象指针
    ref_ptr<orsIProperty> m_inputSourcePtrs;

    //算法参数
    ref_ptr<orsIProperty> m_parameterArgs;
    //输出源名字与对象指针
    ref_ptr<orsIProperty> m_outputSourcePtrs;

    ref_ptr<orsIProperty> m_jobArguments;

    ref_ptr<orsIProperty> m_inputFileNames;
    //ref_ptr<orsIProperty> m_parameterArgs;
    ref_ptr<orsIProperty> m_outputFileNames;
```

```cpp
    ref_ptr<orsIProperty> m_roiArgs;

    orsArray<orsString> m_strInPathNames;
    orsArray<orsString> m_strOutPathNames;
    orsString m_describe;

    ref_ptr<orsIProperty> m_outputMsg;

    ref_ptr<orsIImageSource> m_pAlgSource;

    ors_int32 m_temRectWidth;
    ors_int32 m_temRectHeight;
    orsRect  im_boundingRect;
};
```

ORS_EXE_IMP 宏定义如下：

```cpp
//!!! ORS_EXE_IMP 仅自动产生一个 selfClass 的类,而不会自动注册
//!!! 使用者需要自己注册,才能在插件树中看出来
#define ORS_EXE_IMP(algClass,selfClass,classID,className,describeStr) class self-
Class : public orsIExeImageSourceHelper<orsISimpleExe>,public orsObjectBase \
{
public:\
    selfClass(bool bForRegister): orsIExeImageSourceHelper<orsISimpleExe>(bFor-
        Register)
    {
        if(!bForRegister)
        {
            ref_ptr<algClass> obj = new algClass(bForRegister);
            create(obj.get());\
        }
    }
    virtual ~selfClass(){ }
    ors_string getDesc() const { return describeStr;}
    ORS_OBJECT_IMP2(selfClass,orsISimpleExe,orsIExecute,classID,className);
}
```

10.3.4 界面扩展——GuiExtension

在 MFC 的框架上通过插件扩展菜单、工具条、视图等处理功能。

```
interface orsIGuiExtension : public orsIExtension
{
public:
    virtual bool create(orsIFrameWnd * frameWnd) = 0;
    virtual bool pluginMenu(orsIFrameWnd * frameWnd) = 0;

    virtual BOOL OnCmdMsg(UINT nID, int nCode, void * pExtra, AFX_CMDHANDLERINFO *
        pHandlerInfo) = 0;
    virtual LRESULT windowProc(HWND hWnd, UINT message, WPARAM wParam, LPARAM lParam)
        =0;

public:
    ORS_INTERFACE_DEF(orsIExtension, "gui")
};
```

10.4 一个最小的分布式处理算法软件与客户端构成

图 10-4 和图 10-5 给出了一个 OpenRS 中最小的分布式并行处理的配置例子。其中，OpenRS_Cloud 任务管理器是并行处理的核心，起到任务管理和调度的角色，OpenRS_Cloud 执行器则扮演了处理节点的角色。开发者只需要开发支持 ROI 的插件（参看 8.4.3 节和 10.3 节），就可以利用 ersExeRunner 进行并行处理。

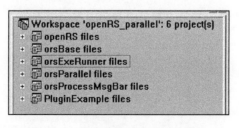

图 10-4 最小的并行处理相关模块

在软件的部署上，由图 10-5 可以看出，要进行并行处理的算法插件（即可执行对象插件）需要在客户端、任务管理器和执行节点上都可见和执行。在客户端，插件对象用于运行参数的提取;在任务管理器中用于提取要执行的子任务;在执行节点上用于子任务执行和结果合并。

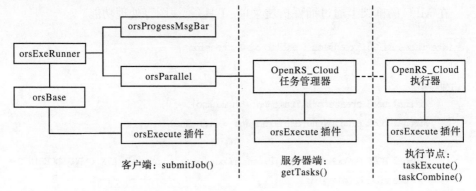

图 10-5　并行处理相关模块联系图

10.5　不同公司或部门软件集成部署方式

不同承担单位(公司)都在软件开发上有自己的积累,形成了各自的遥感数据处理软件。要在一个全新的架构下进行集成,形成具有分布式并行处理功能的平台和系统,是一个重大的挑战。积累既是一种财富,也是一种包袱,继承和发展才是可行的方式。因此,本课题的集成主要在 orsIExecute 层次进行,而不是推翻重来。这样,可以继承使用各单位的原有支持库和核心算法。问题是不能把各单位的动态库简单地复制到一个目录下。如图 10-6 所示,在插件的集成上,不是直接把插件 DLL 复制到插件目录,而是在插件目录下建立插件 DLL 的链接(快捷方式)。

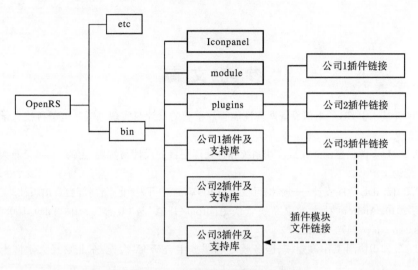

图 10-6 OpenRS 软件即插件部署方式

参 考 文 献

昌燕. 2006. 基于 J2EE 的电信设备网管性能管理系统的设计与实现. 成都:电子科技大学硕士学位论文.

陈镜许. 2011. 多源遥感信息集成应用系统综合处理平台的设计与实现. 上海:上海交通大学硕士学位论文.

董彦卿. 2012. IDL 程序设计——数据可视化与 ENVI 二次开发. 北京:高等教育出版社.

高焕堂. 2010. Android 技术专家高焕堂谈 Android"开源"与"兼容". http://dev. 10086. cn/news/interview/2470. html.

高明. 2013. 知识协同工作流建模、服务规划和服务组合研究. 大连:东北财经大学博士学位论文.

呙维. 2014. OpenRS 开源软件用于提高研究生协同创新能力的方法探索. 科技导刊(中旬刊),10:42-43.

蒋波涛. 2008. 插件式 GIS 应用框架的设计与实现. 北京:电子工业出版社.

江恒彪,关鸿亮,曹天景. 2009. WorldView-2 核线影像立体测图研究. 测绘通报,(5):32-34.

李红臣,史美林. 2003. 工作流模型及其形式化描述. 计算机学报,26(11):1456-1463.

李曙歌. 2006. 基于面向对象知识表示的专家系统的实现. 济南:山东大学.

梁冰. 2012. 基于 RCP 的 FCA 原型系统 XDCKS 的设计与实现. 西安:西安电子科技大学硕士学位论文.

刘建明. 2012. RDF 图的语义相似性度量方法研究. 大连:大连海事大学硕士学位论文.

刘昇,呙维,江万寿,等. 2009. 一种基于云计算模型的遥感处理服务模式研究与实现. 计算机应用研究,26(9):3428-3431.

彭辅权. 2012. Hadoop 集群技术的优化与应用研究. 杭州:浙江大学.

孙步阳. 2009. 机载激光雷达航带拼接技术研究. 武汉:中国地质大学(武汉)硕士学位论文.

孙小涓,雷斌,程兆运,等. 2012. 遥感数据处理运行控制中的工作流应用. 计算机工程,38(4):28-30.

谈小生,葛成辉. 1995. 太阳角的计算方法及其在遥感中的应用. 国土资源遥感,(2):48-57.

唐游. 2012. 海量卫星遥感数据元数据存储系统的设计与实现. 北京:北京邮电大学硕士学位论文.

王炳忠. 1999. 太阳辐射计算讲座. 太阳能,(3):8-9.

汪雷. 2014. 东煤勘探局内网网络协议分析器设计与实现. 长春:吉林大学硕士学位论文.

王淑艳,苏志刚,吴仁彪. 2011. 一种改进的大斜视小孔径 SAR 距离多普勒成像法. 现代雷达,33(9):25-30.

吴亮,杨凌云,尹艳斌. 2006. 基于插件技术的 GIS 应用框架的研究与实现. 地球科学,5:609-614.

邢诚,刘冠兰. 2008. 像素工厂的研究与探讨. 计算机与数字工程,36(9):131-134.

闫志贵. 2010. 基于 Eclipse 的嵌入式调试软件的研究与实现. 武汉:武汉理工大学硕士学位论文.

张立新,魏桐,初洪珍,等. 2014. 依据开放—封闭原则对试验方法代码重构心得. 工程与试验, 54(3):45-47.

张谦,贾永红. 2010. 基于平台/插件软件架构的多源遥感影像融合系统设计. 遥感技术与应用, 25(3):394-398.

张祥, 2008. 面向接口编程详解. http://www.cnblogs.com/leoo2sk/archive/2008/04/10/1146447.html.

张晓. 2007. 面向服务架构中基于语义图服务组合的研究. 济南:山东大学硕士学位论文.

祝若鑫,刘阳,程见桥,等. 2015. 基于云计算的空间矢量数据并行处理. 测绘通报,(3):44-48.

Clugston D. 2005. Member Function Pointers and the Fastest Possible C++ Delegates. http://www.codeproject.com/KB/cpp/FastDelegate.aspx

Golpayegani N, Halem M. 2009. Cloud computing for satellite data processing on high end computer clusters. 2009 IEEE International Conference on Cloud Computing, 88-92.

Guo W, Gong J Y, Jiang W S, et al. 2010. OpenRS-Cloud: a remote sensing image processing platform based on cloud computing environment. Science China (Technological Sciences), S1: 221-230.

附录 A OpenRS 宏定义与模板

A.1 接口定义宏

该宏定义一个最基本的接口,包括引用计数、接口 ID 获取、接口名列表获取、接口查询等接口。其中,getID 实现接口 ID 获取功能。该函数把父接口的 ID 字符串和本接口的 ID 字符串连接,构成层次化的对象 ID。

```
#define ORS_INTERFACE_DEF(parentInteface, interfaceID) public:\
virtual void addRef() = 0;\
virtual void release() = 0;\
virtual ors_string getID() const { return ors_string(parentInteface::getID() + _T
    (".") + interfaceID);}\
virtual void* queryInterface(const orsChar *className) const = 0;\
virtual orsArray<ors_string> getInterfaceNames() const = 0;
```

该宏重复定义了 addRef、release 等纯虚函数。这是因为对象都是继承自 orsIObject,若不重复定义纯虚函数,则会有编译的模糊性问题。

A.2 对象实现宏

该宏对应于接口宏定义,实现接口宏定义中纯虚函数,包括引用计数函数、对象 ID 连接函数和对象名返回函数。根据接口的层数,定义了不同版本的对象实现宏。

A.2.1 无主接口的定义和实现

适用于直接继承在 orsIObject 对象的对象,即主接口为 orsIObject,无需明确指明。

1. 无主接口的对象定义

```
#define ORS_OBJECT_DEF_NORMAL_NO_Primary(objID, objName) public: \
```

```
virtual void addRef(){internalAddRef();} \
virtual void release(){internalRelease();} \
virtual ors_string getID() const {return ors_string(orsIObject::getID() + _T
    (".") + objID);}\
virtual ors_string getName() const {return ors_string(objName);}
```

其中,行末的"\"表示字符串续行,即多行构成一个完整的字符串。

2. 无主接口的对象实现

除定义了对象外,还实现了 orsIObject 的虚表。

```
#define ORS_OBJECT_IMP0(selftype, objID ,objName) \
    ORS_OBJECT_DEF_NORMAL_NO_Primary(objID, objName) \
    ORS_BEGIN_VTABLE_MAP(selftype) \
        ORS_INTERFACE_ENTRY(orsIObject) \
    ORS_END_VTABLE_MAP
```

A.2.2 带有主接口的定义和实现

1. 带有主接口的对象定义

```
#define ORS_OBJECT_DEF_NORMAL(primaryInteface, objID, objName) public: \
    virtual void addRef() {internalAddRef();} \
    virtual void release(){internalRelease();} \
    virtual ors_string getID() const {return ors_string(primaryInteface::getID()
        + _T(".") + objID);}\
    virtual ors_string getName() const {return ors_string(objName);}
```

2. 有主接口的对象实现

除定义了对象外,还实现了 orsIObject 和 primaryInteface 的虚表。

```
#define ORS_OBJECT_IMP1(selftype,primaryInteface, objID, objName) \
    ORS_OBJECT_DEF_NORMAL(primaryInteface, objID,objName) \
    ORS_BEGIN_VTABLE_MAP(selftype) \
        ORS_INTERFACE_ENTRY(orsIObject) \
        ORS_INTERFACE_ENTRY(primaryInteface) \
    ORS_END_VTABLE_MAP
```

3. 除主接口外,还有一个附属接口的对象实现

除定义了对象外,还实现了 orsIObject、primaryInteface、interface2 的虚表。

```
#define ORS_OBJECT_IMP2(selftype,primaryInteface,interface2, objID, objName)
    ORS_OBJECT_DEF_NORMAL(primaryInteface, objID,objName) \
    ORS_BEGIN_VTABLE_MAP(selftype) \
        ORS_INTERFACE_ENTRY(orsIObject) \
        ORS_INTERFACE_ENTRY(primaryInteface) \
        ORS_INTERFACE_ENTRY(interface2) \
    ORS_END_VTABLE_MAP
```

4. 除主接口外,还有两个附属接口的对象实现

与 ORS_OBJECT_IMP2 类似。

```
#define ORS_OBJECT_IMP3(selftype,primaryInteface, interface2, interface3, objID, objName) \
    ORS_OBJECT_DEF_NORMAL(primaryInteface, objID, objName)
    ORS_BEGIN_VTABLE_MAP(selftype) \
        ORS_INTERFACE_ENTRY(orsIObject) \
        ORS_INTERFACE_ENTRY(primaryInteface) \
        ORS_INTERFACE_ENTRY(interface2) \
        ORS_INTERFACE_ENTRY(interface3) \
    ORS_END_VTABLE_MAP
```

ORS_OBJECT_IMP*n* 形式宏主要为自动实现 queryInterface 方法,他们的不同在于其实现类的实现接口的层次与个数相关。若实现类为菱形继承,则系列接口不能使用。

A.3 插件注册宏

插件注册宏 ORS_REGISTER_PLUGIN 主要用于暴露插件输出函数 orsGetPluginInstance()。OpenRS 平台靠 orsGetPluginInstance() 来识别动态库是否为一个 OpenRS 插件。同时,该宏为平台查询插件编译时的平台版本提供了便利函数 orsGetPlatformMajorVersion()。若插件版本和平台版本不一致,则该插件不会被加载。

另外,该宏还为插件提供便利函数 getPlatform() 等。

```
#define ORS_REGISTER_PLUGIN(pluginclass) \
static pluginclass g_pluginInstance; \
static orsIPlatform * g_platformInstance = NULL;
extern "C" PLUGIN_API orsIPlugin * orsGetPluginInstance(orsIPlatform * platformInstance)\{g_platformInstance = platformInstance;return &g_pluginInstance;}
extern "C" PLUGIN_API const char * orsGetPlatformMajorVersion(){return ORS_PLATFORM_VERSION;}\
orsIPlatform * getPlatform(){return g_platformInstance;};
```

附录 B OpenRS 常用模板

为了实现上的方便，OpenRS 定义了一系列模板，用于减少重复代码的编写。

B.1 ref_ptr 的定义

引用计数宏 ref_ptr 用于持有一个 OpenRS 对象。ref_ptr 定义的是一个自动变量，将在变量的生命周期内持有一个对象。使用该宏，要求包装的对象类必须具有 addRef()、release() 接口或方法。在 OpenRS 里面，继承 orsIObject 的对象都可以使用 ref_ptr 进行持有。

```cpp
template<typename _T>
class ref_ptr
{
public:
    typedef _T element_type;

    ref_ptr() :m_ptr(NULL) {}
    ref_ptr(_T* _T1):m_ptr(_T1) { if (m_ptr) m_ptr->addRef(); }
    ref_ptr(const ref_ptr& rp):m_ptr(rp.m_ptr) { if (m_ptr) m_ptr->addRef(); }
    ~ref_ptr() {if (m_ptr) m_ptr->release(); m_ptr=0; }

    inline ref_ptr& operator = (const ref_ptr& rp)
    {
        if (m_ptr==rp.m_ptr) return *this;

        _T* tmpPtr = m_ptr;

        m_ptr = rp.m_ptr;
        if (m_ptr) m_ptr->addRef();
        if (tmpPtr) tmpPtr->release();

        return *this;
```

```cpp
}

inline ref_ptr& operator = (_T* ptr)
{
    if (m_ptr==ptr) return *this;

    _T* tmpPtr = m_ptr;
    m_ptr = ptr;

    if (m_ptr) m_ptr->addRef();
    if (tmpPtr) tmpPtr->release();

    return *this;
}

//ref-ptr 指针的比较操作符
inline bool operator == (const ref_ptr& rp) const { return (m_ptr==rp.m_ptr); }
inline bool operator != (const ref_ptr& rp) const { return (m_ptr!=rp.m_ptr); }
inline bool operator < (const ref_ptr& rp) const { return (m_ptr<rp.m_ptr); }
inline bool operator > (const ref_ptr& rp) const { return (m_ptr>rp.m_ptr); }

//const _T* 的比较操作符
inline bool operator == (const _T* ptr) const { return (m_ptr==ptr); }
inline bool operator != (const _T* ptr) const { return (m_ptr!=ptr); }
inline bool operator < (const _T* ptr) const { return (m_ptr<ptr); }
inline bool operator > (const _T* ptr) const { return (m_ptr>ptr); }


inline _T& operator *()   { return *m_ptr; }
inline const _T& operator *() const { return *m_ptr; }
inline _T* operator->() { return m_ptr; }
inline const _T* operator->() const   { return m_ptr; }
inline bool operator!() const{ return NULL == m_ptr; }
inline bool valid() const{ return NULL != m_ptr; }
inline _T* get() { return m_ptr; }
inline const _T* get() const { return m_ptr; }


private:
```

```
    _T * m_ptr;
};
```

B.2 影像链节点对象帮助模板

为了减少为每个对象重复写同样的代码,定义了影像链节点(影像源对象)帮助模板。

```
template〈typename _T〉
class orsIImageSourceHelper : public _T
{
protected:
    //输入源,不能写成 std::vector〈ref_ptr〈orsIImageSource〉〉, 因为〉〉是流操作符
    std::vector〈ref_ptr〈orsIImageSource〉 〉 m_inputSources;
    ref_ptr〈orsIImageFeature〉 m_pImageFeature;

public:
    doublehelp_zoomRate;
    ors_uinthelp_zoomOut;

    ors_uint help_setZoomRate(double zoomRate)
    {
        if(zoomRate == help_zoomRate)
            return help_zoomOut;

        if(zoomRate > 1.0){
            help_zoomOut = 1;
            zoomRate = 1.0;
            return 1.0;
        }

        help_zoomOut = 1;

        while(help_zoomOut * zoomRate < 1)
            help_zoomOut *= 2;
        help_zoomRate = zoomRate;
```

```
    return help_zoomOut;
}

//ImageSource 接口实现
virtual ors_uint setZoomRate(double zoomRate)
{
    return help_setZoomRate(zoomRate);
}

void setDefaultImageDataActor(ref_ptr<orsIImageData> &imgData,
    orsRect_i &rect, double zoomRate, orsBandSet &bandSet)
{
    if(imgData == NULL)
    {
        orsIImageService *pService = getImageService();
        imgData = pService->CreateImageData();

        imgData->create(getOutputDataType(0), rect, bandSet);
    }

    help_setZoomRate(zoomRate);

    zoomRate = help_zoomRate;
    zoomRate = help_zoomRate;

    /////////////////还原为原始分辨率上的范围//////////////
    double Xmin, Xmax;
    double Ymin, Ymax;
    Xmin = rect.m_xmin / zoomRate; Xmax = rect.m_xmax / zoomRate;
    Ymin = rect.m_ymin / zoomRate; Ymax = rect.m_ymax / zoomRate;

    ////////////////计算在缩小影像上的范围/////////////////
    orsRect_i zRect;

    zRect.m_xmin = Xmin / help_zoomOut;
    zRect.m_xmax = Xmax / help_zoomOut;
    zRect.m_ymin = Ymin / help_zoomOut;
    zRect.m_ymax = Ymax / help_zoomOut;
```

```
        imgData->setRange(zRect, bandSet);
}

virtual bool connect(orsIConnectableObject * object)
{
    if(!this->canConnect(object))
        return false;

    m_inputSources.push_back((orsIImageSource *)object->queryInterface
        ("orsIImageSource"));

    return true;
}

virtual bool disconnect(orsIConnectableObject * object)
{
    std::vector<ref_ptr<orsIImageSource> >::iterator iter;

    orsIImageSource * input = (orsIImageSource *)object->queryInterface
        ("orsIImageSource");

    unsigned int i;
    for(i=0; i<m_inputSources.size(); i++)
    {
        if(m_inputSources[i].get() == input){
            iter = m_inputSources.begin();
            iter += i;
            m_inputSources.erase(iter);
            return true;
        }
    }

    getPlatform()->getLastErrorService()->setErrorInfo(0,"没有链接过该对象");

    return false;
}
```

```cpp
virtual void disconnectAll()
{
    m_inputSources.clear();
}

//遍历接口
virtual ors_int32 getNumberOfInput()  {return m_inputSources.size(); }

//通过索引得到输入链接对象
virtual orsIConnectableObject * getInputObjectByIndex(unsigned index)
{
    if( index < m_inputSources.size())
        return m_inputSources[index].get();

    return NULL;
}

virtual bool setLogicObject(orsIObject * obj) {return false; }

public:

virtual orsDataTYPE getOutputDataType(int iBand) const
{
    assert(m_inputSources.size() > 0);
    return m_inputSources[0]->getOutputDataType(iBand);
}

virtual const orsChar * getFilePath() const
{
    assert(m_inputSources.size() > 0);
    return m_inputSources[0]->getFilePath();
}

virtual ors_uint32 getWidth() const
{
    orsRect_i rect = getBoundingRect();
    return rect.m_xmax - rect.m_xmin;//+1;
}
```

```cpp
virtual ors_uint32 getHeight() const
{
    orsRect_i rect = getBoundingRect();
    return rect.m_ymax - rect.m_ymin;//+1;
}

virtual ors_uint32 getTileWidth() const
{
    assert(m_inputSources.size() > 0);
    return m_inputSources[0]->getTileWidth();
}

virtual ors_uint32 getTileHeight()   const
{
    assert(m_inputSources.size() > 0);
    return m_inputSources[0]->getTileHeight();
}

//virtual double getNoDataValue(ors_uint band, bool *bSuccess)
virtual double getNullSampleValue(ors_uint band)
{
    assert(m_inputSources.size() > 0);
    return m_inputSources[0]->getNullSampleValue(band);
}

virtual double getMinSampleValue(ors_uint band)
{
    assert(m_inputSources.size() > 0);
    return m_inputSources[0]->getMinSampleValue(band);
}

virtual double getMaxSampleValue(ors_uint band)
{
    assert(m_inputSources.size() > 0);
    return m_inputSources[0]->getMaxSampleValue(band);
}
```

//输入波段是输入对象的输出波段数
virtual ors_uint getNumberOfInputBands() const
{
　　assert(m_inputSources.size() > 0);
　　return ((m_inputSources.size() > 0)? m_inputSources[0]->getNumberOfOutputBands():0);
}

//输入波段是输入对象的输出波段数
virtual ors_uint getNumberOfOutputBands() const
{
　　assert(m_inputSources.size() > 0);
　　return ((m_inputSources.size() > 0)? m_inputSources[0]->getNumberOfOutputBands():0);
}

virtual orsDataTYPE getInputDataType(int iBand) const
{
　　assert(m_inputSources.size() > 0);
　　return ((m_inputSources.size() > 0)? m_inputSources[0]->getOutputDataType(iBand):ORS_DT_UnKNOWN);
}

//影像的范围,宽和高
virtual orsRect_i getBoundingRect(ors_uint band=0) const
{
　　assert(m_inputSources.size() > 0);

　　orsRect_i nullrect;
　　return ((m_inputSources.size() > 0)? m_inputSources[0]->getBoundingRect(band):nullrect);
}

virtual orsIImageGeometry *GetImageGeometry()
{
　　assert(m_inputSources.size() > 0);
　　return ((m_inputSources.size() > 0)? m_inputSources[0]->GetImageGeometry():NULL);

```cpp
    }
    virtual orsString getBandName(int iBand) const
    {
        assert(m_inputSources.size() > 0);
        return ((m_inputSources.size() > 0)? m_inputSources[0]->getBandName
            (iBand):NULL);
    }

    virtual orsString getBandType(int iBand) const
    {
        assert(m_inputSources.size() > 0);
        return ((m_inputSources.size() > 0)? m_inputSources[0]->getBandType
            (iBand):NULL);
    }

    virtual const orsIBandMetaData *getBandMetaData(int iBand)
    {
        assert(m_inputSources.size() > 0);
        return ((m_inputSources.size() > 0)? m_inputSources[0]->getBandMetaData
            (iBand):NULL);
    }

    virtual const orsColorTABLE *getColorTable(int iBand)
    {
        assert(m_inputSources.size() > 0);
        return ((m_inputSources.size() > 0)? m_inputSources[0]->getColorTable
            (iBand):NULL);
    }

public:
    virtual orsIImageFeature *getImageFeature()
    {
        if(NULL == m_pImageFeature.get())
        {
            m_pImageFeature = ORS_CREATE_OBJECT(orsIImageFeature, IMAGE_FEA-
                TURE);

            if (m_pImageFeature.get())
```

```cpp
            m_pImageFeature.get()->setImageSource((orsIImageSource *)
                this);
        }

        return m_pImageFeature.get();
    }
};

template<typename _T>
class orsIImageSourceHelper_prop:public _T
{
protected:
    //输入源,不能写成 std::vector<ref_ptr<orsIImageSource>>, 因为>>是流操作符
    std::vector<ref_ptr<orsIImageSource> > m_inputSources;
    ref_ptr<orsIImageFeature> m_pImageFeature;
public:
    double      help_zoomRate;
    ors_uint    help_zoomOut;

    ors_uint help_setZoomRate(double zoomRate)
    {
        if(zoomRate == help_zoomRate)
            return help_zoomOut;

        if(zoomRate > 1.0){
            help_zoomOut = 1;
            zoomRate = 1.0;
            help_zoomRate = zoomRate;
            return 1.0;
        }

        help_zoomOut = 1;
        while(help_zoomOut * zoomRate < 1)
            help_zoomOut *= 2;
        help_zoomRate = zoomRate;

        return help_zoomOut;
    }
```

```cpp
//ImageSource 接口实现
virtual ors_uint setZoomRate(double zoomRate)
{
    return help_setZoomRate(zoomRate);
}


void setDefaultImageDataActor(ref_ptr<orsIImageData> &imgData,
    orsRect_i &rect, double zoomRate, orsBandSet &bandSet)
{
    if(imgData == NULL)
    {
        orsIImageService * pService = getImageService();
        imgData = pService->CreateImageData();
        imgData->create(getOutputDataType(0), rect, bandSet);
    }

    help_setZoomRate(zoomRate);
    zoomRate = help_zoomRate;
    zoomRate = help_zoomRate;

    /////////////////////还原为原始分辨率上的范围/////////////////////
    double Xmin, Xmax;
    double Ymin, Ymax;
    Xmin = rect.m_xmin / zoomRate; Xmax = rect.m_xmax / zoomRate;
    Ymin = rect.m_ymin / zoomRate; Ymax = rect.m_ymax / zoomRate;

    /////////////////////计算在缩小影像上的范围/////////////////////
    orsRect_i zRect;

    zRect.m_xmin = Xmin / help_zoomOut;
    zRect.m_xmax = Xmax / help_zoomOut;
    zRect.m_ymin = Ymin / help_zoomOut;
    zRect.m_ymax = Ymax / help_zoomOut;

    imgData->setRange(zRect, bandSet);
}
```

```cpp
virtual bool connect(orsIConnectableObject * object)
{
    if(!this->canConnect(object))
        return false;

    m_inputSources.push_back((orsIImageSource *)object->queryInterface
        ("orsIImageSource"));

    return true;
}

virtual bool disconnect(orsIConnectableObject * object)
{
    //std::vector<ref_ptr<orsIImageSource> >::const_iterator iter;
    std::vector<ref_ptr<orsIImageSource> >::iterator iter;

    orsIImageSource * input = (orsIImageSource *)object->queryInterface
        ("orsIImageSource");

    unsigned int i;
    for(i=0; i<m_inputSources.size(); i++)
    {
        if(m_inputSources[i].get() == input){
            iter = m_inputSources.begin();
            iter += i;
            m_inputSources.erase(iter);
            return true;
        }
    }

    getPlatform()->getLastErrorService()->setErrorInfo(0, _T("No connec-
        tion to this object"));

    return false;
}

virtual void disconnectAll()
{
```

```cpp
        m_inputSources.clear();
    }

    //遍历接口
    virtual ors_int32 getNumberOfInput() {return m_inputSources.size(); }
    //通过索引得到输入链接对象
    virtual orsIConnectableObject * getInputObjectByIndex(unsigned index)
    {
        if(index < m_inputSources.size())
            return m_inputSources[index].get();

        return NULL;
    }
    virtual bool setLogicObject(orsIObject * obj) {return false; }

public:
    virtual orsDataTYPE getOutputDataType(int iBand)  const
    {
        assert(m_inputSources.size() > 0);
        return m_inputSources[0]->getOutputDataType(iBand);
    }

    virtual const orsChar * getFilePath()  const
    {
        assert(m_inputSources.size() > 0);
        return m_inputSources[0]->getFilePath();
    }

    virtual ors_uint32 getWidth() const
    {
        orsRect_i rect = getBoundingRect();
        return rect.m_xmax - rect.m_xmin +1;
    }

    virtual ors_uint32 getHeight() const
    {
        orsRect_i rect = getBoundingRect();
        return rect.m_ymax - rect.m_ymin +1;
```

```cpp
}

virtual ors_uint32 getTileWidth() const
{
    assert(m_inputSources.size() > 0);
    return m_inputSources[0]->getTileWidth();
}

virtual ors_uint32 getTileHeight() const
{
    assert(m_inputSources.size() > 0);
    return m_inputSources[0]->getTileHeight();
}

//virtual double getNoDataValue(ors_uint band, bool * bSuccess)
virtual double getNullSampleValue(ors_uint band)
{
    assert(m_inputSources.size() > 0);
    return m_inputSources[0]->getNullSampleValue(band);
}

virtual double getMinSampleValue(ors_uint band)
{
    assert(m_inputSources.size() > 0);
    return m_inputSources[0]->getMinSampleValue(band);
}

virtual double getMaxSampleValue(ors_uint band)
{
    assert(m_inputSources.size() > 0);
    return m_inputSources[0]->getMaxSampleValue(band);
}

//输入波段是输入对象的输出波段数
virtual ors_uint getNumberOfInputBands()  const
{
    assert(m_inputSources.size() > 0);
    return ((m_inputSources.size() > 0)? m_inputSources[0]->getNumberOfOut-
```

```cpp
        putBands():0);
}

//输入波段是输入对象的输出波段数
virtual ors_uint getNumberOfOutputBands()  const
{
    assert(m_inputSources.size() > 0);
    return ((m_inputSources.size() > 0)? m_inputSources[0]->getNumberOfOut-
        putBands():0);
}

virtual orsDataTYPE getInputDataType(int iBand)  const
{
    assert(m_inputSources.size() > 0);
    return ((m_inputSources.size() > 0)? m_inputSources[0]->getOutputData-
        Type(iBand):ORS_DT_UnKNOWN);
}


//影像的范围,宽和高
virtual orsRect_i getBoundingRect(ors_uint band=0)  const
{
    assert(m_inputSources.size() > 0);

    orsRect_i nullrect;
    return ((m_inputSources.size() > 0)? m_inputSources[0]->getBoundingRect
        (band):nullrect);
}

virtual orsIImageGeometry *GetImageGeometry()
{
    assert(m_inputSources.size() > 0);
    return ((m_inputSources.size() > 0)? m_inputSources[0]->GetImageGeometry
        ():NULL);
}


virtual orsString getBandName(int iBand)  const
{
    assert(m_inputSources.size() > 0);
```

```cpp
        return ((m_inputSources.size() > 0)? m_inputSources[0]->getBandName
            (iBand):NULL);
    }

    virtual orsString getBandType(int iBand)  const
    {
        assert(m_inputSources.size() > 0);
        return ((m_inputSources.size() > 0)? m_inputSources[0]->getBandType
            (iBand):NULL);
    }

    virtual const orsIBandMetaData *getBandMetaData(int iBand)
    {
        assert(m_inputSources.size() > 0);
        return ((m_inputSources.size() > 0)? m_inputSources[0]->getBandMetaData
            (iBand):NULL);
    }

    virtual const orsColorTABLE *getColorTable(int iBand)
    {
        assert(m_inputSources.size() > 0);
        return ((m_inputSources.size() > 0)? m_inputSources[0]->getColorTable
            (iBand):NULL);
    }

public:
    virtual orsIImageFeature *getImageFeature()
    {
        if (NULL == m_pImageFeature.get())
        {
            m_pImageFeature = ORS_CREATE_OBJECT(orsIImageFeature, IMAGE_FEA-
                TURE);

            if (m_pImageFeature.get())
                m_pImageFeature.get()->setImageSource((orsIImageSource *)
                    this);
        }
```

```cpp
        return m_pImageFeature.get();
    }

protected:  //禁止生成实例
    orsIImageSourceHelper_prop(bool bForRegister)
    {
        if(!bForRegister) {
            m_sourceArguments = getPlatform()->createProperty();

            m_inputSourcePtrs = getPlatform()->createProperty();
            m_parameterArgs = getPlatform()->createProperty();
            m_outputSourcePtrs = getPlatform()->createProperty();

            m_sourceArguments->addAttr(ORS_SOURCES_INPUT, m_inputSourcePtrs);
            m_sourceArguments->addAttr(PARAMETER_ARGS, m_parameterArgs);
            m_sourceArguments->addAttr(ORS_SOURCES_OUTPUT, m_outputSourcePtrs);
        }
    }

    virtual const orsIProperty *getProperty() const
    {
        return m_sourceArguments.get();
    }

    //输入参数信息
    virtual bool initFromProperty(orsIProperty *property)
    {
        ref_ptr<orsIProperty> inputSourcePtrs;
        ref_ptr<orsIProperty> parameterArgs;
        ref_ptr<orsIProperty> outputSourcePtrs;

        property->getAttr(ORS_SOURCES_INPUT, inputSourcePtrs);
        property->getAttr(PARAMETER_ARGS, parameterArgs);
        property->getAttr(ORS_SOURCES_OUTPUT, outputSourcePtrs);

        //兼容旧的调用方式
        if(NULL == inputSourcePtrs.get())
```

```cpp
            m_inputSourcePtrs = property;
        if(NULL == parameterArgs.get())
            m_parameterArgs = property;
        if(NULL == outputSourcePtrs.get())
            m_outputSourcePtrs = property;

        if(inputSourcePtrs.get() != m_inputSourcePtrs.get())
            m_inputSourcePtrs->update(inputSourcePtrs.get());

        if(parameterArgs.get() != m_parameterArgs.get())
            m_parameterArgs->update(parameterArgs.get());

        if(outputSourcePtrs.get() != m_outputSourcePtrs.get())
            m_outputSourcePtrs->update(outputSourcePtrs.get());

        return this->setArguments(
            m_inputSourcePtrs.get(),
            m_parameterArgs.get(),
            m_outputSourcePtrs.get());
    }

    virtual bool setArguments(orsIProperty * inputSourcePtrs, orsIProperty * param-
        eterArgs, orsIProperty * outputSourcePtrs) = 0;

protected:
    ref_ptr<orsIProperty> m_sourceArguments;

    //输入源名字与对象指针
    ref_ptr<orsIProperty> m_inputSourcePtrs;

    //算法参数
    ref_ptr<orsIProperty> m_parameterArgs;
    //输出源名字与对象指针
    ref_ptr<orsIProperty> m_outputSourcePtrs;

};
```

附录 C OpenRS 编译环境与运行环境

C.1 OpenRS 编译环境

C.1.1 目录结构

为了便于不同版本的编译、链接和运行，OpenRS 采用按工程文件（如 *.dsp）（build 目录）、公用". h"文件（include 目录），工程源码文件（src 目录）、库文件（". lib"）（lib 目录）、debug 二进制文件（. exe，. dll 等）（debug 目录）、release 二进制文件（. exe，. dll 等）（release 目录）的方式进行目录组织。设计的目录包括 build、include、src、debug、release、etc。

在 build 目录、debug 目录、release 目录下按 Linux、VC60、VC80、VC90、VC100、VC110、VC120 等子目录的方式分别存放不同编译器版本的工程文件和

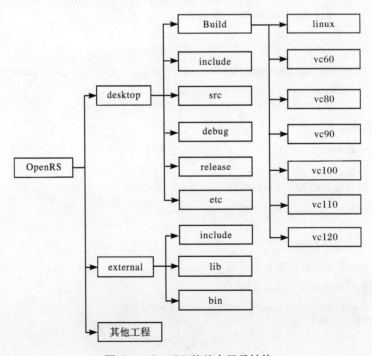

图 C-1 OpenRS 的基本目录结构

编译后的目标文件(图 C-1)。各编译器版本的总工程文件放在该版本的目录下，例如，\openrs\desktop\build\vc60\openRS.dsw 是 VC60 版本整个项目的总工程文件。只要打开总工程文件就可以看到该编译器版本下不同模块或插件的工程。

C.1.2　第三方库目录

第三方库及相关文件放在\OpenRS\external 目录下。提供编译时用到的 C++ 头文件和链接时用到的 lib 文件。

（1）include：从第三方库提取的头文件。

（2）lib：存放编译好的 lib 文件。

（3）bin：存放编译好的 dll 文件。

（4）source：用于存放从 zip 解压的源代码，含有兼容 Windows 所进行的 bugs 修正。

（5）zip：第三方库压缩包。

（6）界面库：BCG 等界面库。

1. include 与 lib 目录的设置

建议把第三方库的 include 与 lib 目录指示放在 VC 的用户目录下(图 C-2 和图 C-3)。以 VC60 为例，可以把\OpenRS\External\Inlcude 目录添加到 VC60 的 Options 的 include 文件目录列表中。

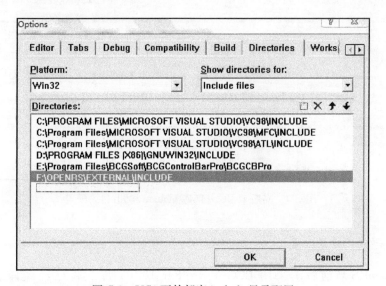

图 C-2　VC6 下外部库 include 目录配置

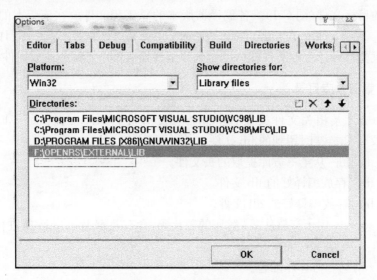

图 C-3　VC6 下外部库 lib 目录配置

对于高版本的 VC，目录设置已经从 options 菜单移除，必须在工程的属性里面设置一次（图 C-4）。在高版本的 VC 中，可以为不同的编译设置不同的 include 目录和 lib 目录。

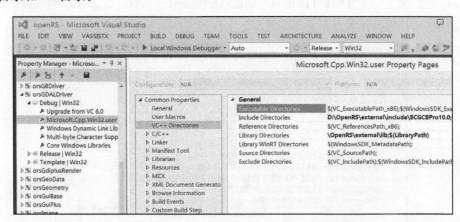

图 C-4　高版本 VC 外部库 include 和 lib 目录配置

2. 界面库目录配置

OpenRS 当前版本采用 BCGControlBar 10.0。库文件已经编译好，放在 external\lib 目录下。头文件已从 BCGControlBar 安装包提取，放在 include\BCGCBPro10.0 下。

C.2 OpenRS 运行环境

C.2.1 桌面运行环境

1. 外部动态库目录

OpenRS 用到的外部动态库，如 gdal，BcgControlBar 等都放在\OpenRS\external\bin 目录下，为了使程序运行的时候能找到这些动态库，必须把\OpenRS\external\bin 添加到 Windows 的环境变量 PATH 中(图 C-5)。

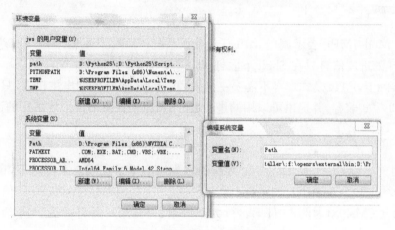

图 C-5 外部库 bin 目录配置

2. 并行调用配置

修改执行程序目录下(orsBase.dll 所在)的 openRS_Parall.conf。该文件配置界面程序提交分布式处理任务时的服务器地址。目前 OpenRS 桌面支持两种模式，即 PTR 模式和 MPI 模式。PTR 模式的配置为

```
[OpenRS Parallel Config File]
MODE: PTR
URL: http://127.0.0.1:18083/
numOfHosts: 1
HOSTS: 127.0.0.1 4
DRIVEMAP: Z:\\192.168.2.175\tmp
```

其中，"Mode:PTR"表示调用 PTR 并行任务管理系统；"URL:http://127.0.0.1:

18083"表示 PTR 的服务调用地址为本机 18083 端口;"NumOfHosts:1"表示主机个数为 1 个;"HOSTS: 127.0.0.1 4"列出主机列表及其提供的线程数;"DRIVEMAP: Z:\\192.168.2.175\tmp"表示 Z 盘映射到\\192.168.2.175\tmp。

MPI 模式的配置为

```
[OpenRS Parallel Config File]
MODE: MPI
MPIMODE: SHELL
numOfHosts: 1
HOSTS: 127.0.0.1 4
DRIVEMAP: Z:\\192.168.2.175\tmp
```

其中,"Mode:MPI"表示调用 MPI 任务分派系统;"MPIMODEL:SHELL"表示调用 MPI 的进度信息只在 Shell 下面打印。

实际上,PTR 模式本身不需要主机列表。PTR 模式下列出主机表,只是为了和 MPI 模式兼容,并提供可以用的进程数计数,以便在界面中限定可用的进程数。

3. 基础数据配置

目前支持 DEM 和 DOM 两类基础数据。在 etc\geoData 目录下,indexedDEMs.txt 和 indexedDOMs.txt 保存了 DEM 和 DOM 的一级编目信息。以 indexedDEMs.txt 为例,文件第一行为二级索引树,第二行开始为分辨率-文件目录索引。

```
2
  90.0 O:\GeoData\SRTM_90m\
1000.0 O:\GeoData\SRTM_1km\
```

具体运用请参考 4.5.1 节。

4. 资源描述目录

OpenRS 基于本体论的思路,通过三元组来描述公有或私有的参数语义。以文件的方式放置在目录\OpenRS\desktop\etc\RDF 中。文件的个数不限。

具体运用请参考 3.4.1 节。

5. 网络服务包装配置

etc\webService 下的 OpenRS_WebService.txt 文件需要自动包装的对象 ID,如

ors.execute.simple.3Danalyst.SurfaceAnalyst.hillshade
ors.execute.simple.BinaryImg
ors.execute.simple.HillshadeClassFuseImage
ors.execute.simple.Statistic
ors.execute.simple.ThreoldImage
ors.execute.simple.imageClassify.supervised.SVM
ors.execute.simple.imageFilter.median

如果该文件不存在，则默认包装所有对象。

C.2.2 分布式处理配置

1. PTR 管理服务器和节点服务器

\OpenRS\ptr\PTRService\ 目录下的 ManagerServerd.exe 和 WorkServerd.exe 分别是分布式并行处理的任务管理器和节点服务器。配置文件在同一目录下，文件名是\OpenRS\ptr\PTRService\ptr.cf。内容和格式为

```
manager.ip              =127.0.0.1
manager.port            =50071
ping.timeout            =3000
ping.timeoutmax         =20000
coincide.num            =2
job.keeptime            =1200000
job.dispatchtime        =10000
machine.checktime       =30000
machine.pingkeeptime    =20000
webservice.port         =18083
my.ip                   =127.0.0.1
workserver.num          =9
agent.path              =f:\openrs\desktop\release\vc60\PTRAgent.dll
```

其中，manager.ip 为任务管理器的 IP 地址，manager.port 为针对节点的网络端口号；webservice.port 为对外服务的端口号；my.ip 为本节点的 IP 地址，workserver.num 为本节点的可分配任务数；agent.path 为 ptr 节点执行代理 dll 的路径，执行代理 dll 应放在 OpenRS 执行程序目录下。

2. monitor

OpenRS 并行处理的状态监控程序 PTRMonitor.exe 放在\OpenRS\ptr\ptr-

Monitor\目录下,可以用于查询可用计算资源和正在运行的作业状态(图 C-6)。

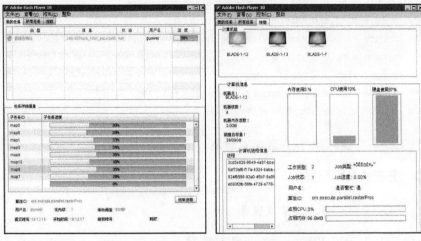

图 C-6 OpenRS 分布式处理监控程序界面

C.2.3 服务网关的安装于配置

假设 OpenRS 安装在 E:\OpenRS 目录,服务程序在 E:\OpenRS\WSDL\C♯目录下,基于 IIS 的 OpenRS 服务配置步骤如下。

(1) 安装 IIS。
(2) 建立网站 OpenRS_Service。
(3) 配置网站:
① 虚拟目录设置,如图 C-7 所示。

图 C-7 OpenRS 服务网站虚拟目录设置

② 默认文档设置，如图 C-8 所示。

图 C-8　OpenRS 服务网站默认文档设置

③ ASP.NET 设置，如图 C-9 所示。

图 C-9　OpenRS 服务网站配置